I0493218

The Water Battery:

Bridging The Fossil Fuel Past With The Clean Energy Future

by Christopher Kinkaid

 Solardyne.com

Published by Solardyne, LLC
Portland, Oregon

ISBN-13: 978-1533138538
ISBN-10: 1533138532

Table of Contents

Preface

"Obsess about the past,
and you may ruin the present,
Forget about the past,
and you may ruin the future"

-Ancient Chinese Proverb-

One of the traditional "arguments" against a Renewable Energy world paradigm has been, "Well, what do you do when the sun goes down? Or, the wind doesn't blow?"

Yes, solar and wind resources are variable. Powerful though they be, this notion of "variability" has been the hook upon which the hat of "Renewables won't work" is often hung.

The "Water Battery" described herein, is a technology which bridges the gap between clean energy technology, and the Fossil-fuel world. The Water Battery is an industrial strength battery technology allowing "variable" clean energy sources like Solar and Wind to be stored safely, then tapped on-demand anytime you wish the power.

The missing link in industrial solar and wind energy has been the "ideal" battery. A battery technology which is economical, powerful, non-toxic, scalable, capable of industrial strength, and doesn't require any exotic materials i.e., no drilling, mining, fracking or strip-mining.

The Fossil-fuel energy paradigm, which has fueled the human industrial economy for the last three centuries has reached a tipping point. For the 21st century, with modern life-styles which are energy-rich in demands, and with over 7 billion people on earth demands a whopping amount of energy. In reality, our civilization requires an enormous amount of industrial energy resources, and will require even more in the future.

The industrial question faced by our civilization now is how do we power the 21st century? How do we fulfill 21st century needs? Do we continue with the Fossil-fuel model? An economic model dictated by variable and high fuel costs, toxicity, and political instability?

The Fossil-fuel world is the ultimate built-in "Have," and "Have-not" paradigm. After all, a very few are controlling something needed by all. Is there a better way?

The Water Battery opens up a new Energy paradigm for the world from a fuel cost, toxic and unfair method of powering our civilization (holes in the ground surrounded by men with guns).   Into an Economic paradigm based on self-sufficiency, no fuel-costs, no toxicity, and no political stress - it works anywhere on earth, and everyone can have one.

The Fossil-fuel paradigm is built on fuel-costs, toxicity, and political instability. The new energy paradigm will be the exact opposite: no fuel-costs, no toxicity, and available to everyone.

The Water Battery is the "missing link" for industrially storing and retrieving electrical energy first produced from variable clean energy sources such as Solar PV and Wind. The Water Battery provides the storage and industrial strengths required by our modern civilization. Being scalable, the Water Battery can be used in residential, commercial and industrial settings. The Water Battery is ideal for mini-grids.

The Water Battery is non-toxic, can be unendingly charged and discharged, can be scaled to industrial strengths, and works in any climate. The Water Battery can be "charged" with any source of electricity, but clean energy generators are the clear choice. Solar Photovoltaic Panels (PV) and Wind Generators are "variable" in their outputs, but have no fuel costs in the energy they produce.

The Water Battery takes this variable Solar and Wind electrical energy and drives an Electrolyser. The Electrolyser electro-chemically disassociates Water into Hydrogen gas, and Oxygen gas which are each stored separately in tanks.

In this Electrical Power Plant electricity is produced at any time on-demand, by recombining the gases in a "Fuel Cell" which delivers the electrical power, and gives you most of the water back. Repeat.

The Water-Battery has no moving parts, no emissions, no toxicity, no soil, air, or water impacts. This battery technology is silent, safe, uses no combustion, no noise, no particulates, no Mercury, no VOCs, no Nitrates, no Sulfates, no radiation. No need for pipelines, no drilling, no mining, no fracking, no railroads, no tankers, none of the "usual" aspects of powering our world. In reality, there is no need to continue using Fossil-fuels - at all.

Indeed, there is a need not too!

The non-toxic Water Battery technology leap-frogs the notion of typical chemical batteries for our industrial needs. Renewable energy has often suffered for lack of a practical economic industrial battery.

Traditional chemical batteries are expensive, often toxic, only work for a limited number of charge/discharge cycles, then need to be replaced at great cost. Further, basing industrial electrical loads on banks of typically small chemical battery cells means many internal connections and high power busbars driving up costs and lowering reliability in large chemical battery banks. With the Water Battery, all of this is avoided.

Imagine, bringing anyone from history into our present time. What would be their reaction to our civilization? Undoubtedly they would be flabbergasted by our modern technology. We can live underwater, fly through the air, travel fantastic distances in hours and be able to talk with people anywhere on the globe in real time using handheld boxes.

Of course, all of these things and many more would be nearly unimaginable to anyone, from any time from the past. Even a few years ago, the Smart Phone was "speculative folly" yet now we communicate worldwide in ways undreamed by most of humanity throughout time.

Indeed, the 20th century just passed was monumental in terms of man's impact on the Earth, and the Earth's reactive impacts on man as our technology expanded exponentially. Man realized, if it can be dreamed - it can be done.

Our visitor from the past would probably be stunned when they ask "so how do you power this amazing world?" And the answer comes back,

"Well, we burn things."

Our guest asking again "where do you get the things you burn?"

"Well, we have holes in the ground, surrounded by men with guns. We have energy markets and we all just duke it out."

I'm sure our guest from the past would be stunned indeed. All of our "miracle" high technology, and we still burn things to boil water - to drive steam engines, or to drive "pistons" up and down in our engines. How we still have our foot in 17th century ideas.

The 18th, 19th and especially 20th centuries transformed the world from a largely agrarian economy into the throws of the Industrial Revolution tapping long ancient photosynthesis in forms of carbon buried long ago and under tremendous heat and pressure.

A Goldilocks situation where ancient biomass has been "cooked" through deep time. Cook it long enough you get Coal. Cook it not-enough and you get shale. Cook it "just right" and you get petroleum, the rock-oil which has been a cornerstone of Industrial civilization for 150 years.

Our Fossil-fuel world has reached an impasse. Continue down the fossil-fuel road and we choke ourselves in our own emissions, and risk the biological viability of earth with all the accumulating toxins being emitting.

Or, we can take advantage of the Water Battery technology and systematically replace nuclear and coal-fired power plants in the world with an "Upgrade" in our energy Operating System.

An "Energy Paradigm 2.0" if you will, with sustainable, clean, and powerful technology which can work for all countries - with no fuel costs, or pollution. An energy upgrade which will work for our children's children. An industrial future which is rich and abundant.

Master inventor Michael Faraday a 150 years ago set in principle the physics of the Water Battery technology. The Water-Battery technology bridges two worlds: the world of our fossil fueled industrial past, with the world of our industrial future based on renewable energy.

The Water-Battery is the non-toxic economical storage technology which makes renewable energies variable inputs into an output which is on-demand 24/7, day and night.

About The Book

This book is about Industrial "Good News" and "Bad News." In short, the Good News is there is a way to power the entire world in a modern energy-rich lifestyle everywhere with no pollution or fuel cost.

The Bad News is if we don't make an Industrial transition from the Hydrocarbon Man into this Solar and Water paradigm, we risk the sudden and rapid collapse of our entire modern civilization.

This book is about the Next Industrial Revolution which is upon the 21st century and the technology which will drive it. The entire evolution of our civilization into a sustainable Industrial Society hinges on what we do, or fail to do in this coming decade. Our modern civilization uses fossil fuels for base-load power generation, transportation, construction, farming fuels, space heating and cooling, feedstock for plastics, fertilizers, and thousands of other purposes in such overwhelming volume this "combustion" now poses an insurmountable toxic threat.

This book discusses a superior energy system not based on fossil fuels, but rather on Sunlight and Water first described over 150 years ago by the Inventor Pioneer Michael Faraday in 1839.

Augustine Mouchot in the 1870s in France applied his technology of using Concentrated Solar energy to produce electricity for decomposing the Water molecule into Hydrogen fuel and Oxygen.

The gases can be recombined in traditional combustion to produce industrial temperatures required by heavy industry. Or, recombining these gases through a Fuel Cell produces electricity on-demand, with most of the water back. Mouchot is demonstrating a truly powerful Industrial Energy architecture based on Sunlight and Water - in 1878. There are many myths which have been perpetrated by the chemical industries for many decades which tend to numb the inquisitive process which would otherwise challenge the Fossil-fuel paradigm. Yet, it must be said, "The Emperor Has No Clothes."

King Coal, and Big Oil run the Industrialized world, and have done so over the previous century. This enormous miss-direction was perpetrated on the world because of three basic facts: energy is valuable, energy is most profitable if controlled, and what better way to subjugate a population (country) than to control its most vital need: energy. The enormous energy concentration of Fossil-fuels made it a valuable item. Indeed, one barrel of oil contains as much energy as 26,000 man-hours of hard labor. If rare coal, and especially oil were controlled by men with guns then a business could be created the world has never seen. And, indeed, after 1892 the die was cast between the "Battle of the Future Fuels" and an energy shake out which would last a hundred years was well underway.

This book challenges the Fossil Fuel paradigm which is based on three pillars of destruction. First, the economic model of charging Fuel Costs where energy is purchased by consumers as a commodity. Second, the high toxicity of burning all fossil-fuels resulting in widespread acid rain, CO_2, mercury, particulates, radiation, and other hazards. And third, the political and economic instability of having the source of a civilization's energy based on "holes in the ground, surrounded by men with guns." The Sunlight and Water Industrial Revolution is described below where technology over 150 years old is resurrected to fulfill the promise and application it held at the very dawn of the modern Industrial age. A choice was made. This is the story of that choice. And as a modern civilization, our choice now.

Chapter One - Energy Empires - As long as Empire's have been on Earth, there has been an "energy crisis." In ancient Greece, Plato lamented in poems how his cherished Forests of Attica in ancient Greece were defoliated as so many people burned all the trees for basic fuels. So that "only the skeleton of the land remains... all of the richer and softer parts have fallen away,... where only the bees can survive." This Chapter explores the long history of "energy crisis" and how an expanding Empire requires foreign sources of fuel, in turn requiring standing armies, and those invaded to respond in kind.

Chapter Two - Heat Engines, And The Fossil-Fuel Age. This chapter explores the birth and use of basic technologies from centuries ago still in wide use today. The Steam Engine is now three centuries old, yet we use basic Steam Engines today for base-load electricity generation from Coal power plants. Modern "nuclear power plants" also boil water driving a steam engine (turbines). The Piston, also three centuries old is commonplace in most cars and automobiles, using an internal combustion architecture little changed in 150 years.

Chapter Three - The Great Energy Wars of the 1890s. Over a hundred years ago Industrial Civilization faced a great battle and shakeout not too different from our present situation. Rockefeller faces his biggest challenge with Tesla's new AC polyphase Power Transmission and Distribution which inspires the Electrical age. Rockefeller's business distilling "kerosine" for Lamp Oil was suddenly obsolete. Paradigms collide as the industrial war over which "fuel of the future" would power the 20th century ensue. Diesel's biofuel burning Compression Ignition engine and Shuman's Solar power plants pumping 6,000 Gallons per Minute all come into play as the great Energy Wars of the 1890s, and the ultimate shake-out which shaped our modern world.

Chapter Four - The Toxic Time Bomb. Pollution and toxicity is building in our world. Measurably increasing in bioaccumulation toxins such as Mercury, Endocrine Disruptors, CO_2, VOCs and others pollutants threaten the very base of the food web. Fossil-fuel use for base-load electricity production, fuel for transportation and heavy equipment, feedstock for multiple products including fertilizer, plastics and other materials have created a toxic "over-load" in the earth's biosphere and hydrosphere.

Chapter Five - Why Sunlight? Why Water? This chapter explores the logic and power of using Solar energy and Water as industrial feedstocks. The sun powers the natural world delivering trillions of kilowatts of power to the earth at any one moment. Driving such immense natural engines including the world's weather direct use of solar energy and indirect use such as Wind power generate energy densities entirely suitable for industrial use.

Chapter Six - Sunlight, and the Water Battery. This chapter discusses the specific water battery technology which converts and safely stores the variable energy production of renewable energy into steady, reliable and available electricity on demand 24/7. The Water-Battery technology is reviewed using Faraday's Electrolyser and Fuel-cell. The Water-Cycle is discussed as an industrial means of storing electricity and releasing electricity at will with no toxicity.

Chapter Seven - The Clean Energy Economy. The Clean Energy paradigm is based on a power supply which is rooted in self-sufficiency. Energy is transformed from a transaction based Commodity model, the Fossil-fuel model, into an economy which generates industrial energy locally. The Solar Water-Battery technology provides a power supply for residential, commercial and industrial loads. The world economy today is largely powered with fossil fuels costing the world economy over $3 Trillion per year.

Chapter Eight - The Thermodynamics of Peace. This chapter explores the thermodynamic requirements of human individuals to live in dignity. Clean soil, food, air and water are essential to healthy human life. To bring the entire population of earth above the poverty level would require more energy than fossil-fuels can provide, nor could fossil-fuels be distributed economically to everyone. A world of human beings living in relative peace will require an enormous power supply. The chapter discusses the practicality of the new Clean Energy paradigm which can obsolete human poverty.

Chapter Nine - Epilogue and the Electric Market Opportunity. This chapter examines the enormous market opportunity the Clean Energy paradigm brings to world labor and capital markets. Labor benefits with this transition working instead of fossil-fuel collection, transportation, and processing into building, operating and commissioning new power plants using the Solar Water-Battery technology.

Introduction

"The disadvantage of men not knowing the past is that they do not know the present."
-G. Chesterton, 1933

It's the best kept industrial secret: you need to pay for energy as a commodity.

Energy is the foundation of every Empire and Civilization known to history. For without basic power for cooking, heating homes, forging metals, and now powering heat-engines for all our industrial needs, how else would civilization be powered?

Fair question, and the answer is extraordinarily essential.

To answer this question, we can look to the past, which this book explores in an effort to understand the path which lead us to where we are today. However, the point is to know the future and calculate the actual power paradigm which would allow access to a world which is energy-rich and available to all people on Earth.

This is in stark contrast to the Energy Paradigm today where most essential energy sources come from remote and scarce holes in the ground, surrounded by men with guns - a precarious way to power a civilization.

Today, energy is a commodity - you need it, you pay someone for it. This economic reality suggests the energy industry runs on the premise "Why teach someone How to Fish, when you can sell them a fish everyday?"

It's the best kept "industrial secret:" that you need to pay someone else for the energy you need to live at any lifestyle you desire. The Sunlight and Water-Battery Industrial Revolution is exactly the point - addressing this "Best Kept Secret." For the entire industrialized world, regardless of politics, East or West, North or South all presently burn copious amounts of Fossil-fuels. Empires need energy.

As a world we burn millions of tons of Coal and over 100 million barrels of OIl - Per day. This enormous daily emission of toxins including ancient CO_2, radiation, Mercury, Nitrates, Sulfates, particulates, VOCs and other pollutants is causing an over loading of Earth's ability to process such toxins. Millions of tons of toxic emission each day is causing a biological impact which threatens the basic life forms which are the base of the food chain for every fish, animal, bird, and human life on Earth. As go the roots, so goes the tree.

Accumulating through three centuries of exponential growth, our Fossil-fuel powered civilization has reached a point where the unimaginable is now apparent. As a civilization we face the breaking point of producing such a toxic impact from our basic consumption of energy we use daily for basic operations such as driving to work, and powering our factories, we've reached a point where a collapse is eminent.

We, as a world, have simply reached a point where the needs of 7 billion people cannot be supported by a power source in use by humans for over 300 years: Fossil-fuels. A power supply of Coal, Oil, and Natural Gas which heretofore has been a profitable and powerful force of nature to be tamed for a very few, and sold to the highest bidder. For what better product could there be than to sell the very energy people need to be civilized?

However, we've pushed this paradigm to such a breaking point that water is becoming un-safe to drink, air un-safe to breath, and foods un-safe to consume from the scorching fertilizers, pesticides, herbicides and genetic modifications "required" to grow a modern crop and the fossil-fuel energy required to produce it. Each Calorie of food energy grown today requires 10 Calories of fossil-fuel energy to produce. Fossil-fuels have become the ultimate business.

And so it is. Since the 1890s, a rapid and energetic pursuit was formed to take advantage of the energy thirst of modern people, with an emerging 20th century to be forged, not on an economy of self-sufficiency, but rather based on the notion of "commodity," that energy should be something bought and sold according to the dictates of a market.

The economic incentive for molding the modern industrial world into a "commodity" energy consumption model - is enormous. The combined global "Energy" market is billions of dollars per day, more than is spent on Healthcare and Education around the world. The Industrial World, it seems, has a power bill. And, it's a whopper.

The actual cost of Fossil-fuels is far more than the apparent market price described as so-many dollars per gallon of gasoline, or dollars per ton of coal. The actual cost is far more in terms of the destruction from toxins produced by burning Fossil-fuels, the economic impacts on farmlands, fields, forests, lakes, streams, rivers and oceans. The destruction of wildlife, and farm animals is widespread resulting from acid rain, mercury, Volatile Organic Compounds (VOC), partially consumed hydrocarbons and other pollutants which are also toxic to man.

The Health Care costs of our toxic civilization endured by many in our populations far exceed the first "Cost" of Fossil-fuels. The environmental costs of Drilling, Mining, Fracking and Strip-Mining destroy local wildlife habitats interrupting life cycles valuable to us, which extend back thousands and millions of years, threatened and damaged from the accumulation of toxicity.

And, sadly, the destruction of our world from burning millions of tons of Fossil-fuels each day, under centuries of increasing volumes, guarantees the Earth faces the extinction of more and more species, at higher and higher rates. Will this terrible progression end with our own human species' extinction?

Unfortunately, if we don't technologically replace Fossil-fuel consumption as the basis or our Industrial Civilization then biological collapse is ensured. Some may argue the timing of collapse, but biological collapse is not just problematic - it's certain.

The proof of this statement is in every indicator you wish to deem important. Mercury levels? Going up. Endocrine Disruptors? Going up. CO2? Radiation? Thermal Pollution? Plastics? CFCs? VOCs? Heavy metals? And, many more - all going up.

There is a point when enough legs of the stool are weakened the stool begins to wobble. Wobbling, always proceeds falling over and this is what we're trying to avoid, yet cannot be avoided if the world continues to power itself by materials dug, scraped, or drilled out of the Earth.

There is a point of saturation. A point beyond which a system cannot further absorb an incoming factor and other reactions, often unintended or desired, begin to express. Fossil-fuels are most often burned. It's combustion. Whenever you burn a

Carbon fuel, like Coal, Oil, Natural Gas, or other Hydrocarbon an enormous amount of CO_2 is produced.

Carbon is relatively heavy, but burn Carbon with Oxygen and you produce CO_2, which is much heavier than Carbon or Oxygen singularly. The heavy mass means for every "weight" of carbon fuels we burn, we produce nearly twice that amount in CO_2. Burning millions of tons per day of Fossil-fuels worldwide produces a global daily emission of nearly twice that amount of CO_2 mass pushed into the environment.

The oceans, and other waterbodies absorb a large portion of this newly introduced Carbon-dioxide and begins to form a weak Carbonic-acid as CO_2 mixes with water. This Carbonic-acid has dire impacts on water quality causing an "acidification" of the water.

Increasing acid levels in oceans, lakes, rivers, streams, and ponds cause an enormous biological stress for crustaceans, phytoplankton, diatoms, algae and other aquatic species which operate at the base of the food web and find it difficult to grow their shells.

If these sensitive life forms are tipped over, then all higher life forms which depend on them: fish, reptiles, birds, wildlife, livestock, and humans will experience a catastrophic failure. A catastrophic failure in soils ability to grow plants, water which is not potable, and air which is full of toxic materials accumulating for centuries causing cancer, disease, and mutating viruses which are rapidly evolving to evade anti-body treatments.

A toxic world is a no-win situation for humanity. A toxic world will collapse the fundamental biological systems upon which our existence is firmly rooted. Without the plant kingdom, we don't eat, and we don't breath. All of the basic nutrition for all higher lifeforms begin with the "autotrophs," the plant kingdom. Push too hard on these sensitive species at the base of our food web, and we're playing with dire consequences.

Oxygen on Earth is produced en-mass each day by solar powered photosynthesis which oxidizes water (releasing Oxygen) and reduces Carbon-dioxide "fixing" and building Carbohydrates and other organic molecules which are the basis of all life on Earth. Photosynthesis is the process which allows almost all life on Earth to exist. Mess too much with vital biological systems, and a grave danger exists especially when those systems are as multi-variable as life systems.

On Earth, there is also a natural Carbon-cycle where Carbon-dioxide produced from volcanism, and the exhaust breath of animals. Processed as a raw material by Plants CO_2 in the atmosphere is being captured through photosynthesis and "reduced" into simple sugars - from which all life is built. When animals or humans consume the plants, or the plants are burned in a fire for example, they emerge from us as a great deal as Carbon-dioxide.

This atmospheric Carbon-dioxide cycle is also "Carbon-Neutral" as the general amount of Carbon-dioxide being released into the atmosphere is being "consumed" by the Plant Kingdom so there are fairly small net-changes in atmospheric Carbon concentrations. This is "Carbon-Neutral."

Humans, however, when burning Fossil-fuels release "Ancient Carbon" which was sequestered by Mother Nature under the ground and "out of the loop" of Carbon-cycling in the atmosphere. Once released, this new Carbon adds to the Carbon-loading of the atmosphere and pushes up the absorption of CO_2 by the oceans- increasing acidification.

Ancient Carbon is only the beginning of what you get when you burn Fossil-fuels. These other toxins begin to accumulate in the environment. It's not just what we consume and burn now, it's all of the Carbon we've been burning for hundreds of years of industrialization.

An enormous cumulative effect builds up. This is the rub. This is the trend which will collapse the basic biological systems of our earth, upon which we totally rely. One of the scariest words in our language: bioaccumulation. The inertia of poisons in our environment and biological systems.

Recent government recommendations for peoples' health call for a reduction in the consumption of deep sea fish to two servings per week. Why? In all of human history we've relied and sustained ourselves with fishing. Why should it be unhealthy to eat more than two servings per week? The answer: mercury poisoning.

Where is all this mercury coming from? Well, it's coming mostly from Coal-fired Power Plants. There is a lot of mercury distributed in coal beds. When we dig up coal and burn it we concentrate and release an enormous amount of mercury spewed into the air and spread over farms, fields, forests, and water ways. All over the world mercury is being emitted in large amounts through our base-load coal-fired electrical power plants, and these mercury emissions accumulate in the environment. Especially, in large deep sea fish.

Water Battery Technology

The old energy paradigm of fuel costs, toxicity and political instability is now poised to be rendered obsolete. The new Electric Power Plant paradigm does not compete with Fossil-fuels - it obsoletes fossil-fuels.

The old model of "have" and "have-note" essential to the Fossil-fuel paradigm will be replaced by an new energy paradigm of no fuel costs, no toxicity, and no political stress available and practical where ever people live.

The 21st century is the greatest opportunity for modern civilization to catch up to our incredible technologies by taking another look at how we generate electricity, how we power our transportation, and manufacturing. The old model only gives us a world of stresses - economic, toxic, and political. The old way is fouling our future.

The Water Battery is the "missing" link needed to bridge two ages: the fossil-fuel age of the past with the Solar-Water age of the future.

Chapter One - Energy Empires

Throughout human history as far back as 10,000 years there has always been Empire. The melting of the last Ice-age brought small wandering bands of hunter-gatherers into the formerly frozen valleys where an incredible transformation of the human experience occurred.

The climate changed and local weather patterns changed by warming up, and groups of humans started "camping" more permanently setting up "bases" of operation to support their hunting expeditions.

An amazing thing happened when humans' started to camp. As humans, being humans, we had garbage-heaps and soon noticed that plants they had harvested from far afield were now growing quite well, in the nutrient rich garbage heap, very close at hand. Bing! A light went on in the minds of ancient peoples; and the benefits of becoming agrarian became immediately apparent.

In some sense, when humans embraced agriculture humanity became "civilized," and at the same moment became "uncivilized." For with the birth of the Agrarian Economy, it wasn't long before bands of men figured out, "Hey, why don't we just make spears and clubs, and wait until our neighbors finish harvesting their crops and just go take it from them?"

And, in this moment the age of Empire was born. When people are raided, they next organized together for mutual defense forming communities, and soon cities. To overcome mutual defense, raiders and attackers would concentrate their forces using organized soldiers to achieve overwhelming strength - and the age of conflict was born. The Age of Empire had now taken root as surely as newly domesticated plants and animals.

A seemingly unending struggle ensued between peoples for resources abroad, when resources at home couldn't keep up with demand. The birth of civilization brought the birth of pirates, empire (organized pirates), and the rest of the population, all jostling for position, power, survival, wealth, or at least a chance at life. Empire requires fuel, have it and you have the world. Have not, and you posses not a step.

The Ancient Greeks

Looking back into history, the ancient Greek civilization is an early example of how Energy and Empire go together and are inseparable. As goes one, so goes the other.

By the 5th Century B.C. ancient Greece was already in an Energy Crisis. Charcoal was the preferred fuel for cooking, heating, and powering kilns and forges.

The big problem was - it takes a lot of wood to make charcoal. To make charcoal large pits were dug. The pits are nearly filled with wood which are set alight. As soon as embers begin to form the pit is covered with earth and allowed to bake.

In a short time the anaerobic conditions under the soil act to Pyrolyse the Carbon in the wood concentrating the Carbon. Charcoal is a concentrated "wood" which doesn't produce as much smoke, and produces higher temperatures than wood for a nice even burn, ideal for indoor home braziers used as heaters.

Making charcoal was the "refining" of the ancient world. As the ancient Greek civilization grew with their seafaring success, their populations flourished and expanded, and with this new success came the need for more and more fuel.

In ancient Greece this meant wood. It was such a problem that soon Plato's beloved forests of Attica were deforested so severely he lamented "all the richer and softer parts have fallen away, ... and the mere skeleton of the land remains," mourning the once lush forests of his youth. Plato also wrote of the extensive erosion of the land "the water which runs off the bare earth into the sea." Such was the extent of the energy crisis and the consequences faced by the ancient Greeks, and the dramatic impacts on their local environment which ensued.

This was the basic challenge of many humans living together. Sources of water, sources of waste-water disposal, sources of food, cooking oil (olive oil) and fuel were of paramount importance. The amount of wood required in the Bronze Age was staggering to fuel local populations. Supporting thousands not only for cooking and heating, but powering the forges of Copper and Tin to supply large numbers of weapons - all required vast volumes of wood. The demand for wood now rapidly exhausting local supplies.

Also, there was the energy intensive production of Triremes, the great ramming war ships which protected the vast seafaring city state of Athens when invaded by Sparta, or worse the Persians. The great war ships of ancient Greece required huge quantities of wood for shipbuilding perpetually increasing in the number of Triremes required. By the 3rd century B.C. it was a death penalty to cut down an olive tree for fuel, so scare and valued was wood and charcoal - protecting the national strategic interest of something so vital as olive oil.

The ancient Greeks found themselves facing an ever bigger problem - they needed more wood, and badly. Once the wood is depleted in local environs and the domestic supply is exhausted, then the only direction left is to look abroad. Unfortunately, abroad usually means someone else's land.

To get at the wood, you'll need a standing army which requires a lot of weapons, and a lot of food, which in turn requires a lot of wood. For an ever increasing energy appetite the ancient Greeks found themselves in a quandary. Expanding empire requires expanding energy sources, in turn requiring an expanding empire.

Of course, those who once owned the wood felt compelled to raise an army of their own in defense, and conflict becomes the name of the game as Empires and all those who lived around them are drawn into a constant unending cycle of struggle and violence. Wood, was the vital fuel of the day. If you're an Empire in the Bronze Age, it was fundamental to your survival, and you did what you had to do to get it. Including defend yourself from those who would invade and take yours from you.

The ancient Greek civilization flourished and expanded through the centuries by responding to their energy crisis not only with war, but with their wits. The ancient Greeks are famous for their natural philosophy and their contemplation of the forces of nature. Seeking the nature of reality from reality itself.

The Greek's many gods are alive and well in the meaning of natural forces turning out well, or not well for mere mortals. However, the ancient Greeks with their keen interest in all the world developed something incredible in response to the practical energy crisis they faced: solar architecture.

The ancient Greek architects gained and expressed incredible knowledge of solar energy and it's practical application for heating and cooling homes and buildings - no small feat when doing so with wood would be impossible or highly expensive at best. No other than the ancient Greeks with their development of true democracy went hand in hand with the quality of life afforded its citizens, (or most of them) demonstrating this heart-felt democracy across class lines - in no small part solar powered.

The very notion of Civilization was highly valued by the ancient Greeks and their use of solar architecture for many of its citizens was a testament of this fundamental Greek value. The expression of Solar architecture gave the ancient Greeks a mastery and partnership with nature (the gods) they highly coveted.

Solar energy was a practical source of energy. The ancient Greeks saw well how the sun gave life to the plants and warmed the earth each Summer to the great benefit and comfort of all. Solar access and good health were synonymous and ancient Greeks instigated the world's first solar access laws as each citizen born has a right to the sun, and all the benefits therein derived.

Moreover, the Greeks observed the sun was high in the sky during the Summer, and hot on the skin. By facing their porticos South and placing eves extending just long enough, then the sun high in the sky during the Summer would be shaded, yet allow the warming sun's rays into the space when the sun angle was low in the sky during Winter. This solar architecture promotes cooling in the summer, and heating during the winter - all with no moving parts, no fuel costs, and no toxicity.

Compare this with the alternative of importing wood, or expensive charcoal. Dealing with moving it when needed and the resulting fumes and smoke. The constant cleaning and other support activities coupled with ever increasing high costs - make the practical use of solar energy: "civilized."

Solar architecture saved the ancient Greek civilization by providing much needed heat and light for homes and buildings on a daily basis greatly relieving the amount of wood otherwise required. Solar architecture developed and flourished in ancient Greece designing urban spaces with East-West and North-South street grid patterns. Streets so arranged provided solar access to all buildings and provided vital solar access for warming homes during the winter and providing solar access for all domiciles and citizens of Greece to "promote good health, and comfort." An amazing aspiration for the ancient world.

The ancient Greek architects knew how to build second stories higher on the Northern end and shorter on the Southern end to maximize the solar gain without shading the Northern part of the building. Greek solar architects and engineers were expert with solar time-keeping using their Gnomons and sun dials. By the position and length of the sun's shadow the hour, and season were well known as well as all daily and seasonal changes.

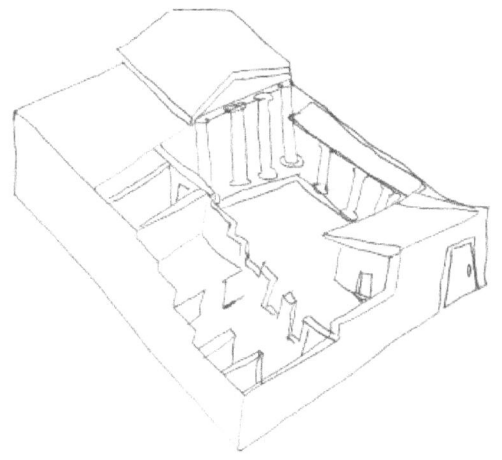

The Classic ancient Greek house was designed rectangularly with low walls facing South with an internal courtyard. The entrance was from the South which entered into the internal courtyard flanked on the left with rooms and to the right by a covered path.

The covered pathway on the East side flanked the courtyard and lead to a Northern Room, or rooms, which open to the South and was warmed and lit by the sun's daily rays. Rooms on the West end of the courtyard would receive the morning sun warming first. Those rooms were typically used early in the day for preparing and consuming meals.

The privacy, utility, intimacy and elegance of an ancient Greek home is a brilliant use of solar architecture to enhance the quality of life for the inhabitants with no fuel cost, or pollution which surely would occur if attempted with wood.

This expression of architecture was the powerful and inspiring relationship civilization has with its natural environment, and how natural energies properly engaged can provide a means of great practical value and worth. Having warm walls and floor bricks in the winter from solar heat-gain is very nice, especially compared to cold bare stones as cold as ice as the alternative.

To the ancient Greeks - solar energy practically applied was a fundamental expression of the value of human life, and the very heart of what it meant to be civilized. Human civilization and it's natural relationship with nature to the ancient Greeks - with solar architecture was the essence of living as a civilized being.

The Ancient Romans

The Roman civilization to emerge soon faced the same problem, and opportunity. By the Second century B.C., most of Italy was already largely deforested, and the sprawling Roman Empire was feeling their own Energy Crisis. Again, it was wood, or the lack of it.

However, the energy crisis for the Roman Empire was many times more severe than experienced by the ancient Greeks. The shear scale of the rapidly expanding Roman Empire, and the wide variety of climates and locals put tremendous strain on local sources of fuel: namely, wood. The Romans, moreover, have entered the Iron Age which required furnaces of high heat. Higher heat requires even more fuel.

The higher the heat produced in a forge, the more exotic the materials can be made, and in the case of smelting Iron it was a breakthrough of epic proportions. The Bronze Age was a huge leap over the Copper Age before. Bronze, an alloy of Copper and Tin is much harder than tools or weapons made from Copper alone.

The leap from Copper to Bronze, thousands of years before, was one of those innovations which immediately changed human civilization and the balance of power between competing peoples. Iron metal was harder as a material, and much easier to acquire than Copper or Tin. Tin, was especially rare and early Bronze Age citizens had to travel great distances to get it. Iron ore, however, was found everywhere and this changed the economics of tools and weapons.

Being much harder than Bronze, Iron tools and weapons were vastly superior and again civilization changes based on how hot you can make a furnace. Now, more than ever, everything depends on an ever expanding supply of wood.

Creating heat in a forge was the key innovation which led to the Iron Age, and to produce great heat you needed lots of fuel. Again, the shortage of wood, and it's by-product charcoal put enormous pressure on the Roman Empire to find new sources of wood. By hook, or by crook, the Roman Empire was on a mission: find new sources of energy and exploit them.

By the Second century B.C., the Roman's were importing wood into Rome from as far away as the Caucausus over 1,000 miles away. Early Roman historians such as Piney the Elder wrote about the fuel shortages of this period. Italian forests were so denuded of trees that traveling the roads along the Italian peninsula made travelers imagine thieves hiding behind every tree trunk along the road.

The demand for fuel by the expanding Roman Empire was an energy crisis in full swing by the 1st century B.C. Wood and charcoal was needed for cooking, heating, lighting, and most importantly for powering forges for Iron weapons and tools needed in ever increasing numbers by the Roman Empire's military industrial complex.

Soon in the common era Rome's energy crisis became so destabilizing, drastic measures were required by the Emperor. In response to shortages on the Italian peninsula, an entire fleet of ships called naviculari lignarii (wood ships) were commissioned to systematically import timber from foreign lands.

The Roman wood ships would scavenge the Western Mediterranean raiding France, Spain and North Africa for sources of wood. Imagine the entire industry required to build a fleet of ships (made from wood) and equip them with crews which can invade foreign lands, fell trees, cut trees, load trees onto ships, and sail them back to major ports in Italy - then go do it again. An enormous enterprise, the

Roman military industrial complex requires vast amounts of wood for ship building supplying an ever increasing military need for Iron weapons and tools required to defend, promote, and execute Roman expansionism.

Harvesting and transporting wood from German, French and other land accessible forests was difficult from a military tactical standpoint. Again, vast number's of soldiers were required to invade, defend, then fell, cut, and process wood for transport over long distances by land. Further, a long thin line of men and carts along a single road in a long thin column exposes the flank to the enemy, prompting an ever increasing need to beef up military escorts.

Barbarians, Gauls, and other invaded populations, themselves pressed by migration from the Asian steppe, would harass and attack long Roman columns on the flank and cause great losses and interruptions of fuel supplies back in Rome. Shortages of fuels would cause great civil unrest, and Roman leaders were hard pressed to bring in more fuel, more bread, and more of everything expected in Roman society.

In effect, the Roman Empire became an expanding "Ponzie" scheme as ever increasing supplies of energy would need to be sourced, supplying an ever expanding Empire and expected lifestyle. An expanding military Empire requiring ever increasing sources of energy. It was a positive-feedback loop which shook the Roman Empire to it's knees.

The enormous "Wood Ship" fleets required to gather wood from around the Mediterranean required great sums of money to build, maintain, crew and operate. The emperors of Rome faced with ever decreasing conquests found new sources of capital by raising taxes.

By the Third century of the common era taxes were so rampant in Rome that many of the intelligencia and successful citizenry began to flee the city to retire to their various country estates - to escape endless tax increases and currency devaluations.

Roman emperors are notorious for rapidly raising taxes, or the outright confiscation of estates if deemed, by whim, "for the good of the state." The rising expense of the "Wood Ships" vital to keep the stability of the empire resulted in a capital flight from Rome with many leading citizens retreating to their remote villas - while they still possessed them. The Roman Emperor's fiscal policy hurt the very productive citizens Rome needed to survive if not prosper.

This exodus of talent and capital-flight from Rome only resulted in a further deterioration of the Roman Empire leading to its breakup and eventual fall in the Fourth century.

The fall of Rome was as much precipitated by an ever expanding and long seeded Energy Crisis as by any other conflagration of factors. When Rome could not expand enough to fuel its internal consumption and military needs it fell, and a civilization that lasted for millennia became no match for the consequences of not having enough fuel.

Roman Solar Innovations

In response to this energy crisis, by the First century B.C., brilliant Roman architects began to borrow and then improve upon the Solar architectural principles pioneered by the ancient Greeks before them - towards the same ends: lowering energy costs and improving the quality of life.

Roman culture was centered squarely on the spoils of empire. All Roman citizens had free access to water, the circus, and the baths. This was a grand accomplishment in the minds of Romans and gave them a sense of belonging and access to the riches generated by the tyranny of empire. The Romans came, they conquered, and they taxed the heck out of anyone under their yolk. The great colosseum funded by the sacking of Jerusalem, for example.

For citizens of Rome plunder was the foodstuff of existence, and a righteous claim to dominion over other peoples, creeds, and races - the natural order of things in the Roman mind. The strong, should subjugate the weak, and thus the spoils of empire flooded in from every corner. However, this required an extensive appetite for energy and long before the fleets of "Wood Ships" Rome would have to apply it's great engineering skills to the task. The Roman Empire is really unparalleled by the incredible engineering feats achieved by Roman artisans, engineers, designers, and architects including the military.

The Roman legions were astounding engineers building all types of roads, bridges, war machines, building construction, water wheels and many other accomplishments of incredible power output. On military campaign the Roman legions built mote and stockade - each night! To protect the soldiers from attack at night each camp built 2 meter deep motes around the camp to make invasion difficult, then a wooden stockade to give enough time for the Roman soldiers to react to a sneak attack.

The Romans devoured wood at every turn. The typical Roman soldier was a skilled artisan adept at carpentry, wood working, construction, with often specialized and overlapping skill sets for flexibility of mission, including of course - fighting.

The Roman soldier was expected to march all day, build their nightly mote and stockade, and do it again day after day for weeks at a time. Enormous military power, reach, and effectiveness through organization and repetition. Military power based on unabated fuel consumption: wood.

The famous Roman Aquaducts are an excellent example of Roman engineering far ahead of the rest of the world. The incredible precision required to slope the Aquaducts from 0.10 to 0.25 percent allowed the water to flow just fast enough. Too fast and the water would spill out, too slow, and the water would not flow in sufficient volume.

The engineering precision of this approach is something to behold and demonstrates a mastery of practicality. Fording rivers and valleys and all changes of topography the Roman arch coupled with hydro engineering delivered millions of gallons of water per Day to the Roman capital with no moving parts, no fuel costs, no pollution or consumptions, and with little maintenance.

The Energy Crisis building up steam in the early Roman empire also manifested in the practical use of natural energy as shown by the earlier Greek empire: solar energy. Real, intense, and practical the Romans applied their inventive genius toward using the sun for practical purposes with spectacular results.

The early Roman empire heated villas, palaces and important public buildings using the "Caldarium." Firing bricks with holes which lined up to form channels when put together allowed a fire box located in the basement to circulate hot air through the walls and floors using these innovative bricks to warm the building.

How sophisticated to have forced air heating for keeping the floors and walls warm two thousand years ago? However, to have central heating was expensive and consumed an amazing amount of wood to keep fueled, and slave labor to operate and keep stoked and maintained. A single villa could easily consume two cords of wood per day to heat the marvelous home, and when wood became scarce, very expensive.

The great Roman architect Vitruvius of the First century B.C., was a great student of the "ancient" Greek empire and was well aware of the writings of Socrates, Aristotle and other Greek architects on solar architectural principles.

Vitruvius, however, built on ancient Greek engineering noting the Roman empire is so expansive there are many climates which must considered when designing the architecture of a house. All the ancient Greek city-states shared a common climate. The Romans, on the other hand were everywhere from the heat of North Africa to the chill of British winters. Designing buildings and houses for a given location the climate should be the first consideration.

Vitruvius advised in the First century B.C.:

"We must begin by taking note of the countries and climates in which homes are to be built if our designs for them are to be correct. One type of house seems appropriate for Egypt, another for Spain, ... one still different for Rome, and so on with lands and countries of varying characteristics. This is because one part of the Earth is directly under the sun's course, another is far away from it, while another lies midway between these two... It is obvious that designs for homes ought to conform to diversities of climate."

Vitruvius went beyond Aristotle and Socrates by selecting which rooms are best placed where in a temperate climate. In extremely hot climates such as North Africa, Vitruvius recommended homes with rooms open to the North to avoid the heat of the day. The dining room should be placed on the Western side to receive the warming sun's rays in the afternoon and early evening adding in comfort and pleasure.

Vitruvius wrote of other innovations in solar heating homes and buildings. Under the dining room floors a central pit was dug and filled with pottery shards and small stones covered finally with black sand. When sunlight entered the windows into the room it would begin to heat the "thermal mass" under the floor which in the early evening would begin to re-radiate a wonderful heat. Even the slaves must have appreciated warm tiles under their feet. Vitruvius comments on the advanced thermodynamics of thermal gain, and how to control this marvelous heat for practical consumption.

One of the greatest Roman inventions in solar energy technology was the Greenhouse. Some inventive Roman around the 1st century B.C. realized windows could be made using glass, or very thin mica, a translucent rock. The windows would keep out the rain and wind, but allow in light and warmth.

The colored glass industry existed in Roman for many centuries before this time. Glass was blown into bulbs, then spun around very fast to stretch the glass into a cylinder. At the right moment this cylinder was cut with iron shears and laid out flat onto an iron sheet covered with sand. The invention of the solar green house effect greatly impacted solar architecture.

The great Roman Baths were in large part solar powered. Not only did the Roman's consider solar heat more healthy it was also a practical way to provide the enormous heat required by the Roman bath lifestyle.

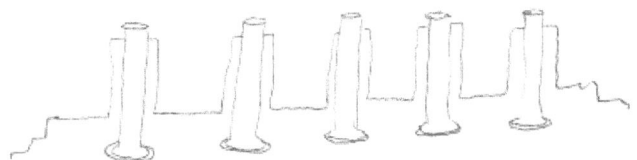

Roman ruins from Pompey shows window openings for 10 foot panes to trap the Sun's heat for Thermal Gain

Energy is fundamental to Empire, it's fundamental to civilization no matter the century, and is presenting a substantial challenge if our civilization is to survive, let alone thrive in the 21st century.

Before we get to the 21st century, let's take a look at the 19th century, for most of our "modern" energy industry is rooted in the 19th century, with technology little changed in the power industry in it's objective: combustion.

Chapter Two - Heat Engines and the Fossil Fuel Age

All Coal-fired, and nuclear power plants all over the world are simple heat-engines. Simple? Well, in terms of basic architecture, Yes. The entire design goal of a modern electrical power plant is essentially the same as it was three hundred years ago: to boil water and power a steam engine.

Today, coal-fired power plants, which account for large amounts of base-load electric power generation around the world, are all based on combustion. We burn things to power our world. We burn a lot of materials (Carbon), a staggering amount by any measure. Burning millions of tons of coal per day, world consumption of coal requires enormous mining operations. Coal mining is not a pretty business and has significant environmental impacts for nature, and for miners. Consider the enormous footprint, mostly hidden from direct view, required to operate a Coal Power Plant scheme.

First, you need the coal, so you mine out the hills. Next, you'll need to transport the coal to remote power plants typically placed away from population centers to hide the emissions. Out of sight, out of mind. Transporting millions of tons of coal per day world-wide requires railroads, and vast networks of track. Coal doesn't travel well. With so many exposed beds of recently crushed rock-coal they shed a significant amount of coal dust during transport. We've delivered our coal to the Power Plant, and now it's offloaded, piled up, and then, of course, burned.

Modern Electric power plants are Steam Engines, and require a "condenser" to cool off the working fluid (steam) back into water to be reheated. To cool these large steam engines copious amounts of water are required. Steam engines have an enormous thirst in their condensers and consume water to be evaporated causing cooling, or after use as a coolant simply returned to the river, or ocean from which it came. Unfortunately, this increases the water temperature of the water body which can have terrible effects on local wildlife species often sensitive to tenths of a degree change.

When water temperatures rise, the water's ability to hold dissolved Oxygen is lessened (Henry's Law) which dramatically impacts fish, crustaceans, algae, and other vital base species. Then, of course, we must add the costs and impacts of the toxins released by burning coal into the air. The incredible filth spewed into the air from burning coal on such a scale is also staggering. The partially consumed hydrocarbons, particulates, radiation, mercury, VOCs, Nitrates, Sulphates, and bulk CO_2 emitted by modern coal plants is pouring ancient carbon into the hydrosphere increasing the toxicity, and energy of the atmosphere.

The increase of energy in the atmosphere and oceans will express itself in more severe weather events, and changing weather patterns causing increased deaths, property damage, loss of economic activity, crops and in general more frequent disruptions of local economies.

The coal-fired power plants around the world are emitting millions of tons per day of CO_2 which is being absorbed in part by the oceans. There is no biological system on earth unaffected by our base-load coal-burning power generation.

Nuclear power plants, being heat-engines, also present thermal pollution, mining impacts, and require enormous amounts of cooling water. Being a steam engine, the modern nuclear power plant has the additional nasty attribute of dealing with materials of such high toxicity (radio active materials), a catastrophic failure would pollute vast tracts of our world rendering them uninhabitable.

The modern nuclear power plant is a high tech way of performing a low tech goal: boiling water and driving a steam-engine. Is there a better way to power an industrial world?

From the birth of the First Industrial revolution 300 years ago nothing compares to the 20th century in terms of man's sudden and dramatic impact on the Earth, and how we power our modern world. Even for the last 10,000 years as man moved from a precarious existence against the cold, harsh and relentless climate of an Ice-age to the sudden warm-up which spring boarded human life with the development of something new: civilization.

Nothing in human history would compare to how far technology has accelerated in the last three centuries, and especially the 20th century. What does the 21st century have in store?

For all of our technological wonders in our modern age, it's amazing how our energy system is actually so primitive in objective. After all, all of our nuclear and coal fired power plants all do one thing: boil water.

The 21st century will see the industrial evolution from a "combustion" model driving steam engines and pistons to a non-combustion model based on renewable electricity and water.

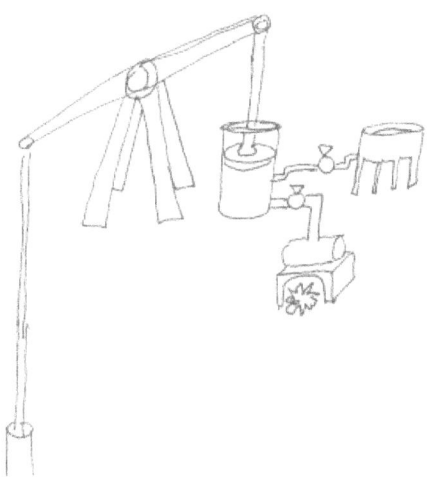

Since 1712, with the introduction of Newcomen's engine we started "boiling water" to perform work and the world turns a corner. Unleashing the incredible potential of the First Industrial Revolution this one machine changes everything. Miners have long fought flooding in the mines, which had been a curse to the early miners eager to reach rich veins of coal, if only they could pump out the flood water.

The Engine itself was by our standards primitive, but at the time revolutionary consisting of a large piston inside a cylinder closed at the bottom, and open at the top attached to a "rocker" arm to power a large pump.

This Engine was an "atmospheric" engine as the work done was achieved by the weight of the atmosphere above the piston. It worked by placing a large fire box and boiler under the piston and opening a valve to allow hot steam to enter the bottom of the cylinder and push the piston up. Keep in mind the "up stroke" is not the power stroke.

As with a hand pump the power stroke is on the down stroke. The up-stroke just repositions the piston, so in this case the steam doesn't do much direct work, but it does allow a great trick. Once the steam was introduced into the cylinder and the piston resumes the Up-position, the steam valve was closed and another opened valve which sprayed cold water into the bottom of the cylinder. This caused a rapid condensation and a partial vacuum is quickly formed. It's this partial vacuum under the piston which allows the atmospheric pressure above the piston to push down and perform the power stroke. To keep this rocking action on the pivot bar connected to the pump the cycle is repeated, closing the cold water valve and reopening the steam valve to return the piston to the up position, and repeat.

This "atmospheric" engine, also called an "external combustion engine" as the combustion is done outside the working cylinder, changed the world by unleashing the notion of burning coal for power, not just for heat. Coal, shale, peat, wood and other naturally sourced fuels were well known to the ancients and provided fuel for heating, lighting, and cooking for thousands of years. The difference here is now coal is being tapped for power - unleashing the notion of burning fuels to produce power in a steam engine - multiplying the power of a man's muscle ten-thousand fold.

By 1750, James Watt invented a fundamental improvement to the Steam Engine by realizing instead of heating up and cooling down the same cylinder taking time in between strokes, why not separate these functions by creating a permanent Boiler (Hot area) which you always keep hot, and a separate Condenser (Cold area) which you constantly keep cool.

Watt's great innovation, among many to follow was to circulate an internal "Working Fluid" in a closed-loop which would enter the boiler and vaporize producing high pressure and doing work, then circulated into a "condenser" to be cooled back down into liquid water, now ready to be introduced into the boiler again, and repeat.

The steady-state conditions of Watt's Boiler and Condenser allowed the working fluid (Water being boiled into Steam in the boiler and returned to liquid water in the Condenser) to be cycled continuously and greatly increased the thermodynamic efficiency (and power) of the basic steam-engine.

James Watt's innovations turned the First Industrial Revolution into high gear. Now, small factories could turn their looms, mills, lathes, and saws anywhere they wanted to set up. Until this time - water wheels were the mainstay power-train for small manufacturers. Basing power on moving water limited the location you can set up a mill, for example, since you needed to be situated correctly to tap into moving water.

The Watt Steam Engine now allowed coal to be the "prime mover" and the fuel of choice in 1750. The result was a rapid expansion of distributed coal-fired steam engines which drove the looms and turned the lathes and mills launching the First Industrial Revolution into exponential growth.

The British Empire in the 18th and 19th centuries expanded industrially all over the globe. You remember, "the sun never set on the British Empire." England's advantage early on was coal seams near the surface. Quickly exhausting the access to easy coal deeper seams would often flood. When mines flooded the Newcomen engine, then Watt's Steam Engine pumped out the mines unleashing a flood of coal from the mines transported by Railroad to markets throughout England. Newcomen's, then Watt's engines launched British industry into the forefront. England took over the world - largely because they figured out how to tap the enormous ancient solar energy stores of coal. England's rapid industrialization, and new found industrial might was not un-noticed by rival Empires. The Germans also had great coal resources, and raced to catch up with England.

France, however, only had coal from its remote frontier such as the Pyrenees bordering France and Spain which was far away from their main industrial centers. Forced to import coal from England and Germany, France lagged far behind in the internal railroad infrastructure required to import coal from their frontiers, and was forced to compete with England, and Germany under great disadvantage.

Solar Technology emerges as an industrial power supply.

To compete with England and Germany, France was forced into being resourceful, and a moment in history opened to a "new way" of finding a "prime-mover" to power commerce. In 1860 a Pioneer emerged which could free France from it's Energy Crisis and lack of coal with a new technology. A technology which as Empires discovered before in history, is potent, effective, absolutely practical with the additional advantages of no fuel costs, or pollution. This new energy source: concentrated solar energy.

Augustine Mouchot was a brilliant Mathematics professor in Lycee de Tours, France, and through the 1860s and 1870s set himself to the enormous potential of solar energy applied to practical application. Mouchot was a Pioneer as he opened the door to demonstrable solar power for purposes which are of great importance for the local wine industry which flourished all around Tours. Mouchot built solar concentrating machines which concentrated solar energy to very high power densities, and set them to work.

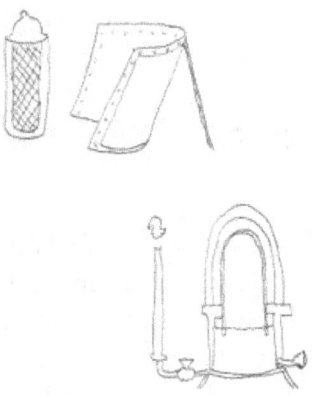

Above - shown Mouchot's early Solar Cooker. Right figure shows Mouchot's early Solar Water Pump. Solar energy heated the air inside a Copper Caldron surrounded by a glass bulb. The heated air expanded pushing down on water inside the bottom of the Caldron forcing the water out a thin pipe.

Mouchot's objective was to power steam-engines with concentrated solar energy, and his machines achieved spectacular results. Starting simply, Mouchot took a Copper cylinder closed at the bottom and sealed at the top with a wooden lid. Around the Copper cylinder he placed a glass bulb spaced about one inch from the Copper. This vertical assembly was placed in the sun where the sun's rays heated up the Copper cylinder trapped by the glass producing a green house effect. Behind the Caldron beyond the shadow he placed a curved vertical reflector. This back reflector irradiated the North side of the Cooker with additional solar energy and this simple use of solar optics unleashed a flurry of solar devices.

Filling the inside of the Copper cylinder with food to cook, Mouchot would routinely cook stews, roast meats, vegetables, even baking bread with his solar oven and touted it as a way all peoples' of the tropical colonies of France perform daily cooking with no need for fuel.

But this was only the beginning for Mouchot. By tweaking his design he increased the performance and capacity and was soon distilling wine into Brandy "with a most agreeable flavor." A most impressive feat to the French who take their wines and brandies most seriously. Mouchot forged on increasing the power density of his solar concentrators, and soon assembled a Solar Water Pump. Mouchot was beginning to adapt his optical collectors for higher and higher concentration ratios.

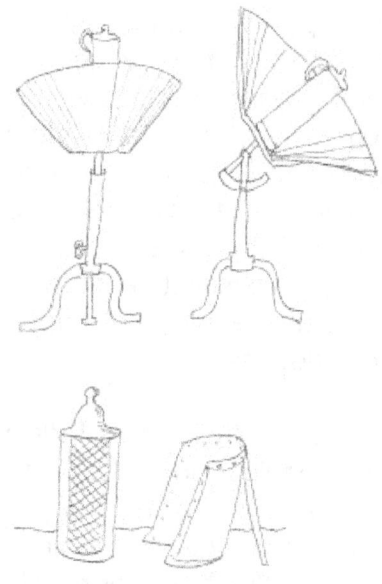

Above -views of Mouchot's different Solar Cooking apparatus using an adjustable Conic reflector and below a vertical parabolic back reflector circa. 1868

Mouchot invented an optical solar concentrator which resembled an inverted lamp shade. The large open mouth was directed toward the sun. The inner walls, sloping inward, are covered with thin silver reflectors. A boiler placed at the focal line at the middle of the device would receive solar radiation from all angles at once and this greatly increased the power density and thermodynamic performance the device achieved.

Now Mouchot was at a full gallup. Increasing the size of this concentrator and changing out different receivers for different purposes Mouchot demonstrated an impressive industrial machine with a wide range of utility.

In an age where coal is already king, Mouchot is offering an energy starved France something better. Power with no fuel costs, and no pollution. An unlimited industrial solar power supply.

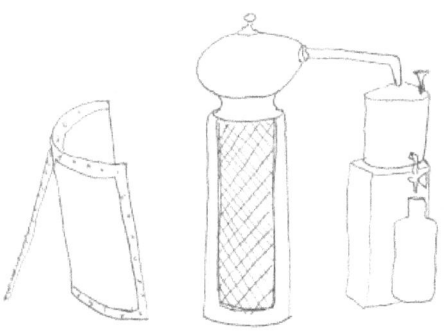

Above - view of Mouchot's early distillation apparatus. Wine, or water in the center caldron surrounded with a glass jacket trapped direct solar energy and additional sunlight from the back reflector shown left causing vapors

inside the vessel to rise and collect near the top forced by internal pressure through a tube through a separate vessel of water to cool the vapors condensing them into liquids.

Mouchot demonstrated his solar engine to the French Academy of Sciences and began to receive some financial backing to continue his research and development. Soon Mouchot was able to Pasteurize wine, distill gallons of wine into Brandy at a batch, and now distill water into potable clean water from brackish and salt water sources.

Mouchot understood that water borne diseases were a big problem for the Tropics, colonies of France, and for all people. Mouchot was able to take any type of water, no matter how brackish, toxic with mineral salts, or sea water itself, and was able to boil this water distilling out clean potable water free of pathogens.

Note: Today, over 5,000 children die each day after a long battle with dehydration due largely to water borne diseases. Why aren't we using Mouchot's devices?

As true in 1868, as it is 2016, people all over the world suffer and die horribly from drinking unclean water. Infants and young children being the most vulnerable. Mouchot fought to have his solar distillers and cookers employed worldwide, and demonstrated his technology in France, and spectacularly in Algiers. It is a modern tragedy, In

2016 an estimated 2.5 billion people still use sticks and dung as basic fuel for heating, and cooking, and so many souls suffer and perish - for want of clean water.

Mouchot's solar concentrators were beginning to offer the world real solutions to real problems. Mouchot receives funding and quickly expands his solar applications. He develops a Solar Water Pump which pumped 500 gallons per Hour for irrigating vineyards. With no fuel costs or pollution, and using the solar energy which happens to fall on his device - he pumped 500 Gallons per Hour an impressive result in France in the late 1860s. Working towards his goal of powering larger and larger steam engines Mouchot innovates another first with solar energy: making ice.

Using Ferdinand Carre's method of refrigeration Mouchot concentrated solar energy to high temperatures. Carre's method used a combination of Ammonia and water. Using solar energy to boil this working fluid the Ammonia evaporates and is directed towards a "condenser" which returned the separated Ammonia to a liquid state. Once this happens the Ammonia pours over some part of the container, containing the food you wish to refrigerate, and presto.

The Ammonia absorbs the "heat" from the Food container cooling the container as the Ammonia evaporates and becomes vapor - pulling the heat from the food container. In another condenser the vapor is relaxed into liquid and remixed with water to returned to the boiler, and repeat.

Mouchot used Carre's method replacing the "boiler" which formally burned coal for heat with his Solar Concentrator and in the heat of the Algerian sun, Mouchot made ice.

Can you imagine being in Algeria under the intense sun in 1868, and this mathematics professor produces ice? Mouchot was the first person in the world to produce ice from sunlight and introduced a technology which can be life changing for so many people.

Above -Mouchot's Solar Engine powering a printing press at Paris exposition winning a Gold Medal

Mouchot was in full swing demonstrating solar cooking, distilling, water purification, ice-machines, water pumps to irrigate the world, and steam-engines with no fuel costs or pollution - then came a disaster.

In 1870 the Political leader of France Napoleon III, declared war on Prussia. Prussia promptly invaded France and Mouchot's funding dried up. Further, Tours was overrun and all of Mouchot's engines, notes, drawings, with the entire contents of his laboratory sacked.

Bad enough his laboratory was destroyed, but strangely all of his prototypes and papers had been removed. Was it the Prussians? Perhaps, Mouchot was becoming famous in Europe for his solar inventions and Prussia didn't want France to gain an advantage in the Colonial wars of the time.

However, Mouchot may have had many enemies. Particularly in the French railroad and mining factions which wanted more government investment in traditional coal power which Britain and Germany still lead, constantly lobbying to forget this solar energy technology which they perceived as a great commercial threat to their business.

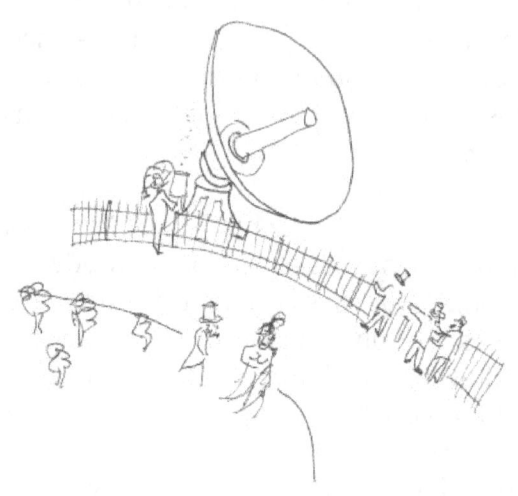

Above- Mouchot's Solar Engine on display in Paris

The Coal producers and Railroad interests needed to bring coal in from the frontier, and wanted all government resources focused on their business under the flag of nationalism.   Mouchot had many enemies from competing commerce. After all, Mouchot was a Mathematics professor, why should such an academic threaten our profitable business? Grumbled the coal companies.

The future was fluid, and no one knew which fuel would ultimately win the Energy Wars. Solar energy suddenly became a very real threat to the Coal producers fearing their motto "Why teach you how to fish, when I can sell you a fish everyday?" could be exposed.

Mouchot, after all, was teaching people how to fish!

After the disastrous years post 1870, Mouchot pulled himself together and built even more impressive machines. Powering larger heat-engines Mouchot's solar concentrator drove water pumps, and other steam-power engines producing 5 horse-power in the sun, an incredible feat in France with no fuel cost, pollution or fuss.

It was in the late 1870s however, which drew Mouchot's true brilliance. Following the work decades earlier by Michael Faraday in London, Mouchot put it all together and stunned the industrial world.

Mouchot proposed to use concentrated solar energy to produce electricity which would drive Faraday's Electrolyser reducing water into Hydrogen gas and Oxygen gas. Mouchot knew you could combust the two gases producing extremely high temperatures ideal for heavy industry, or you could recombine the gasses in another of Faraday's invention, the Fuel Cell, releasing great deals of electricity on demand - plus, you get the water back. Faraday had invented the "flow" battery which as long as the gases flowed into the fuel cell - electricity came out.

Imagine in a time when Solar Photovoltaic (PV) panels and modules, commonplace today didn't exist. Indeed, it would be decades until Einstein in

1905 published the photoelectric effect before the physics of solar PV panels were even understood.

Mouchot envisioned using another recent invention developed years earlier by Seebeck: the thermoelectric pile. The thermoelectric power device worked by soldering together two dissimilar metal plates, such as Iron and Copper. Once soldered you heat the junction between the metals and electricity is produced across the junction. Mouchot arranged dozens of Seebeck's thermoelectric junctions replacing his boiler and under his solar concentrator produced a "respectable current." Mouchot then used this solar powered electrical current to "drive" Faraday's electrolyser producing Hydrogen fuel from water.

Here we are in 1878, and Mouchot is producing rocket fuel in his laboratory from solar energy and water. Non-toxic rocket fuel. Amazing. But, there is something even more incredible.

Ever heard much of Mouchot? Such an industrial accomplishment, as pioneering as the first Flight at Kitty Hawk, but where is it discussed? Where is it described? How many science books, or current engineering students in the United States of any discipline are even vaguely aware of Mouchot? Or, Faraday's Electrolyser and Fuel Cell? No doubt, engineering students are aware of fuel cells, but without the electrolyser the whole significance of the Water-Cycle is lost.

Note: The author finds the nearly universal lack of reference to Mouchot's work throughout the modern science curriculum not a coincidence, or an idle omission. It's such an important story, the author must consider why anyone would want this story omitted from modern education?

Suggesting, and indeed demonstrating a solar power generation technology with no fuel costs, toxicity, or political stress capable of producing rocket fuel industrial strength - would be in direct conflict with the Coal, Oil and Natural Gas business interests who want to sell you a fish every day. As true one hundred years ago, as it is today. After all, do you expect the Fossil fuel industry to just, roll-over? Giving up a multi-trillion dollar industry? Not likely.

The global Fossil-fuel business combined exceeds $6 Trillion dollars each year. That's more money than most countries GDP. If you were in command of a $6 Trillion dollar business, would you be telling school children for the last hundred years about solar powered Electrolyser's and Fuel Cells?

Would you tell four generations of school children the high costs, military conflicts, toxic air, soil, water and food from Fossil-fuels they suffer could all be replaced with Solar Energy and Water? Would any generation who knew about this, tolerate a continuing polluted hostile Fossil-fuel world? Will ours?

Human Nature

There is a famous study where people of three villages who live under a dam were interviewed. The villagers who lived the furthest downstream were asked "what do you think about the dam?" They would scream "Oh, the dam, if it breaks our children, our houses, our lives will all be swept away! It's alarming!"

Next, the villagers who lived closer to the dam were asked "what do you think about the dam?" They replied, "Well, the dam is an issue, could be dangerous if it broke."

Finally, the study interviewed the villagers which lived directly beneath the dam, "what do you think about the dam?" They answered, "what dam?"

Sometimes, when human beings are near an extreme danger they may initially react in fear, but very quickly humans "block out" the danger and get on with business. Makes sense really, imagine you're an early human and you're being attacked by a Saber Toothed tiger. Before the attack, human reaction may be fear a very useful tool in detecting danger, but humans to survive can't dwell too long on impending danger when faced with attack, they must act and turn their attention.

In the case of the Saber Tooth tiger we have Fight, or Flight, not standing there wide-eyed, frozen, only to become lunch. In the case of living under the dam, loosing one's sense of fear to a prolonged threat, may back-fire leaving us oblivious to the danger.

Ignoring the danger, utterly and just tending to our daily business can make us numb to reality.

Sometimes, human's react when living too close to a looming catastrophic danger, by blocking out this danger from daily thought so as not to be overrun with emotion. This behavior runs the risk of healthy fear-impulses being essentially "forgotten."

Regarding Fossil-fuels, it appears we live too close to the dam. We're so culturally used to using Fossil-fuels, it's so engrained in our daily routine that imagining a world doing something different is hardly considered.

Too close to the dam

Mouchot's Solar powered Water-cycle power system is almost never reviewed, discussed, or even contemplated for many years in the scientific or anthropological literature. For all of the Freedom we enjoy as citizens, for example, how easily our world views can be sabotaged by commercial interests firmly rooted in which research is funded at the College, University and Federal levels, and which is not. Which basic industrial information is suppressed, omitted, and most damaging of all: ignored. There is no better weapon of suppression than indifference.

Is there one example of anyone, or group in a University, Government, or Institutional organization experimenting or building a solar powered water-hydrogen power system? The author challenges the dear reader to find even one. Coincidence? Not if you want to keep your job!, we'll get back to that one.

Would you ever expect the Fossil-fuel interests to allow even the slightest research to be performed at any foundation, institute, college or university which would expose their industrial scheme? I challenge readers to find one Federally funded, institution or foundation funded, or University research project which emulates anything near what Mouchot was demonstrating in 1878. Not one? Doesn't that seem a little - funny?

The Emperor has no clothes

Empires have their interests. Since the 1890s, it's become a fossil fuel powered world. Not that it wasn't before, after all, coal powered the First Industrial Revolution starting 300 years ago. However, the dawn of the 20th century, brought a Second Industrial revolution which could have gone several different ways. A hundred years ago a biofuel, or solar power paradigm was demonstrable at the threshold of practicality, and was a serious threat to the fossil-fuel paradigm.

The transition from the "External Combustion Engines" of the First Industrial revolution powered by coal, now enters a Second wave of the "Internal Combustion Engine" and the Energy Wars commenced between the Oil Industry, Tesla's new AC Electrical Power Grid, Rudolph Diesel's Biofuel burning Compression Engine, and the Solar Pioneers - all coming to a loggerhead. Which energy paradigm rule the 20th century?

The winners and losers of this shakeout for the "Fuel of the Future" would set the course of powering human civilization for the next hundred years. The 20th century, a triumph in technological advances also spawned so many world wars we started numbering them. More hundreds of millions of humans suffered, and died in the 20th century from poverty, famine, abuse, disease, and war then the previous dozens of centuries combined.

Our modern civilization was birthed at great cost, and the heights of our success and innovation is surely measured by the depths of our collective sorrows. With all of the freedoms our modern technology offers, we are still held captive in 2016 by the economic and environmental chains of Fossil-fuels established in the 1890s, and centuries before.

The Energy Wars of the 1890s didn't end until 1913 where an ultimate shakeout left only one player standing: John D. Rockefeller.

A catastrophe of epic proportions occurred which shaped and doomed the 20th century to come: World War I.

Chapter Three: The Energy Wars of the 1890s

In the 1890s an all out war erupted between the Energy interests and competing paradigms who's outcome would define the next hundred years. The battles which ensue are between the major energy players as the world comes to grips with the dawn of a new century rapidly approaching.

How will this new century be powered? Who will control this market? Which technology will win the day and capture the ultimate prize? Who would win this supreme shakeout?

The energy industry has ruled the world of empire and expansion for millennia, and now it has the 20th century in its sights.

The country and interests who controlled energy could control their own destiny, and deny enemies theirs. An energy paradigm based on "Why teach you how to fish, when I can sell you a fish every day?" was the most profitable, most controllable, must exclusive, and most efficient in defining the world as "have," and "have-not" this condition being the engine of profit. Chemical Fossil-fuel companies and certain countries with these resources begin landing squarely on the side of "haves." The entire world of energy consumers: the have-nots.

The Energy Wars of this period are significant today because the same concerns were expressed about how a new century which promotes peace, not conflict could be powered?

Imagine the 1890s, the amazing and exciting world was expanding on all fronts. Miracles in technology from telephones, to street cars to motion pictures sprouting everywhere. None more spectacular worldwide than the Chicago Worlds Fair of 1892-93. For at this fair the world changed. Tesla's polyphase AC current Electrical Power Transmission and Distribution system wowed the world by lighting up the entire fair with electric lights!

Lighting today is just flipping a switch, but to those living in the 19th century, and All humans before back into deep time - lighting was a pain in the backside.

A Scottish Chemist James Young in the 1850s cracked the fractioning of rock-oils into distillates. Before 1859, when the rock-oil industry burst on the scene with Edwin Drake's deep rock drilling the act of lighting the room was a major enterprise and took a lot of work, time and money. Throughout antiquity indoor lighting was always problematic, insufficient, smelly, dirty, and expensive.

Throughout history lighting was mostly achieved in homes and palaces by burning animal fats, plant oils, waxes, whale oils, and other flammables typically at great cost and inconvenience.

The modern Oil Industry was founded distilling "Kerosine" from rock-oil, and after 1859, the world of lighting changed fast. The market was lamp oil - often called illuminants. Internal combustion engines pioneered by Otto, and Karl Benz didn't come on the scene until the late 1880s and didn't exist at the dawn of the Oil Age. For the Oil

Industry, "in the beginning" meant producing Kerosine for lamp oil.

When Rockefeller entered the Kerosine Rock-Oil Business there were thousands of different oil "wildcatters" drilling for oil and hundreds of "refineries" processing any number of different grades of "Kerosine." The Lamp Oil market was hot and Rockefeller wanted in but what made Rockefeller so special?

John D. Rockefeller made some radical business decisions which leap-frogged everyone. Seeing the end-game, Rockefeller focused on Refining instead of the very risky business of drilling. Let other's take the risk of "wildcatting" which had great returns or total busts, and instead focused on the bottle-neck in the production stream: refining. Further, Rockefeller realized the market was over-saturated with refineries. At the time Rockefeller launched into the rock-oil business, "everyone and their uncle" was in the business without any standards for quality which varied greatly. Rockefeller realizes an inspiration and decides to make the same can, the same color, the same quality (refining process) of Lighting Fluid worldwide, in effect, the Standard Oil. And it worked. Whether you bought a can of Kerosine in Beijing or New York it was the same product, it was the standard upon which you could rely. It's Standard Oil.

After decades consolidating his empire of Kerosine refining and distribution, Rockefeller, going full speed turns into 1892 and hits a wall. Electricity is here - and it obsoletes Rockefeller in a single blow. The Chicago World's Fair and Exposition of 1892-93 powered and brightly lit with Tesla's new polyphase electric AC power transmission impressed if not flabbergasted everyone - worldwide.

The entire fair was spectacularly lit and Tesla's polyphase AC current brought the wonder of electricity to everyone. Flip a switch - and light instantly comes on - it's bright - no smoke - no fuss - amazing, and with AC power transmission could be distributed everywhere.

Now come the fireworks. John D. Rockefeller had just spent the last 25 years building the largest Kerosine and therefore Oil business in the world. The business was refining and selling rock-oil at tremendous profit. In comes the Chicago World's fair with Tesla's lighting system and everything blows up in Rockefeller's face.

Electricity!

John D. Rockefeller, the conqueror of hundreds of competitors, the consolidator of refining capacity throughout the country and abroad, the father and chief architect of the world's largest chemical company - dashed! By something called "E-l-e-c-t-r-c-i-t-y." What?!

Imagine being in Rockefeller's board room when the world got ahold of electricity - and in one fell swoop threatened decades of Rockefeller's empire building. To the world over the last 30 years, Kerosine was an incredible revolution over any lighting technology which came before. But, electricity now trumped it all. And with the flip of a switch. The Kerosine business was doomed.

After all, what can compare to the convenience of pushing a button - and instantly light appears?

Rockefeller, was having a fit. However, seeing the writing on the wall Rockefeller had a breakthrough and insight which only occur when such men's souls are stretched to their limits and catastrophe has no way out but victory! He saw a way to turn the tables, much as he has always done in business, by staking out a set objective, then turn on his advisories and bring all tubes to bear.

Rockefeller would transition from lighting fluids, to Power production by fueling internal combustion engines with a distillate they used to dump in the rivers at night as a waste product: gasoline. And, just in time for a revolutionary new machine: the automobile powered by the internal combustion engine. Rockefeller from the clutches of utter defeat, pulls a rabbit out of his hat - and finds a new world to conquer.

As if the rapid emergence of electricity wasn't enough to shake up Rockefeller, the experience was so traumatic if left Rockefeller with a new hardened purpose: never to be caught so exposed again.

Rockefeller doubled his efforts and approached this new emerging market for his distillates with an even sharper focus and iron will.

By early 1905, automobiles with internal combustion engines were themselves having a shakeout from a very popular competitor - the Electric Car. And soon, by the more powerful Diesel engine ideal for trucks rapidly coming on the scene which can run on nearly any Biofuel, including animal fats, and plant oils sourced from corn, soybean, and peanuts.

Rudolf Diesel, in the later part of his life became the world's greatest advocate for biofuels, and its most credible. After all, Diesel had an engine which ran on compression ignition which allowed a wide range of fuel sources including beneficial Nitrogen crops especially peanut oils to power the engine.

"The use of vegetable oils for engine fuels may seem insignificant today, but such oils may become, in the course of time, as important as petroleum, and coal tar products of the present time." Rudolf Diesel - 1912.

In a world of energy scarcity, regardless of its composition and source, and in the shadow of the constantly waring countries of Europe, the specter of a World Energy Paradigm based on countries growing their own fuels to relieve economic and political stress - was greatly welcome. And, important to Diesel the military adventurism and struggles for energy resources based on fossil-fuels - could be greatly reduced.

Rockefeller understood the implications for his business should this "Diesel" engine be fueled by Biofuels instead of his rock-oil distillates. To Rockefeller - Rudolph Diesel's biofuels paradigm was as mortal threat to his rock-oil empire as disruptive as Electricity proved earlier.

Diesel was enormously credible being already world-famous for his engine designs, and because he brought the engine technology which allowed everyone to grow their own industrial fuels. Rudolph Diesel in the mind of Rockefeller posed yet another major threat to Rockefeller's iron grip on rock-oil distillate markets.

Who Killed Rudolph Diesel?

On the eve of World War I in 1913 Rudolph Diesel was taking a ferry across the English Channel on his way from Europe to London to sign a lucrative deal with the British Navy to outfit English U-Boats with Diesel's new powerful marine Diesel engines.

Diesel's compression ignition needed no "spark-ignition" as did internal combustion engines burning gasoline. This was a significant advantage for boats, ships, and shipping. Spark-ignition using spark-plugs was very dangerous in the confines of a ship. Gasoline, was a very volatile fuel at sea with constant movement hazardous vapors and air could easily mix and explode.

Diesel's compression engine technology was far safer and produced more power than Gasoline internal combustion spark-plugs - ideal for heavy loads such as Marine applications, including U-boats and military vessels.

A mystery occurs aboard ship with Diesel's disappearance late one night having dined earlier with friends. Returning to their respective cabins Diesel is never seen again. Biographers claim that Diesel committed suicide, by jumping overboard, they cite "he had debts." Well, most industrialists carried debt, so the author doesn't see enough motivation from this corner. Indeed, Diesel was traveling to England to sign some large contracts with the British Navy guaranteed to bring in cash.

No, something else happened. The night Diesel disappeared over the side of the ferry there were many of the "usual suspects." The ultra German nationalists at the time certainly would have held a motive to kill Diesel, as they didn't want anymore technology transfer to the British as war was nearing. However, Diesel was a German national hero, valuable before and during any conflict.

Regarding technology transfer, it seems to the author in essence the "cat is already out of the bag." Interfering with this particular licensing agreement certainly wouldn't stop the British Navy from acquiring the Marine Diesel Engine technology they sought through "knock-offs" or third parties.

It seems reasonable the German intelligence services were probably "tailing" Dr. Diesel, however the author doesn't see enough motivation for the German militants to have Diesel killed. After all, if the militants did want to murder Diesel why do it at relatively high expense aboard ship on the confines of a ferry where concealment would be difficult? Why not assassinate him using "normal" methods as was routine? In the morning after his disappearance at sea, it's reported that Diesel's hat and overcoat were found neatly folded under a deck chair.

This fact, leads the author to suspect a great deal of care was taken to "hide" this murder, with some effort to imply suicide. It was reported hundreds of miles away from the ferry route a body was recovered floating in the waters of the North Sea, a few days after Diesel's disappearance with personal effects later identified (by his son) to be of Rudolph Diesels. However, the body was buried at sea by the fishing trawler which found it floating so far from the crime scene, and thus the body was never positively identified or examined for foul play - and no criminal investigation was ever raised. The case was quickly ruled a suicide and closed.

All of these factors suggest the question - who had the most to gain from Diesel's death? The author's research suggests Diesel was murdered and a ruse of suicide planted. The person most to gain with the "elimination" of Rudolf Diesel was John D. Rockefeller - he had the motive, and through his operatives the means.

The author suggests Rockefeller took Diesel's message of biofuels and industrial self-sufficiency as the next extreme threat to his hard-fought empire. Having just recovered a decade earlier from the entire Kerosine market collapse and replaced by electricity, Rockefeller was not willing to risk another displacement. Rockefeller knew Diesel was right about biofuels. The author suggests that's why Diesel was murdered by Rockefeller's henchmen, and why the melodramatic move of neatly folding Diesel's coat and hat under the deck chair was employed.

In any event, the loss of Rudolf Diesel was a deathblow to the world of renewable biofuels. With the launching of World War I just months after Diesel's death, the world was thrown into war and the Petrol-chemical companies cleaned up. If you're in the "industrial fuels" business, there is nothing more profitable than war.

There was a brief moment in time just at the dawn of the 20th century where energy Paradigms could have been different. However, Rockefeller was at the top of his game, and his well deserved reputation for brutality through his henchmen set the stage for the 20th century. Energy empire was alive and well. The "fuels" required for civilization would not be based on "people learning how to fish for themselves," it will be based on "I'll sell you (and everyone else) a fish everyday."

No stranger to government contracts, Rockefeller found war the most profitable enterprise yet - if you're selling fuel. Wars' use a lot of fuel and material, and with the prosecution of the World War I, John D. Rockefeller made fortunes hand over fist.

In the crucial years leading up to 1913, before WWI could consolidate his global empire Rockefeller had yet another threat across his bow. Mouchot's work with solar concentrators in France thirty years earlier had not gone un-noticed in the states. A brilliant Swedish inventor and engineer John Ericsson took on the cause of solar energy powering the world.

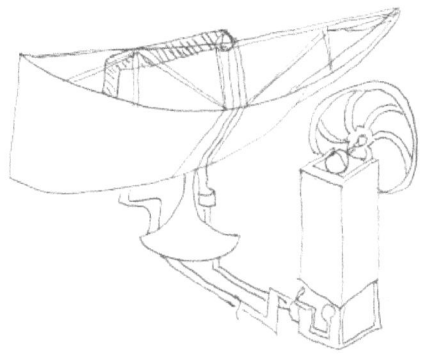

Shown Above - Ericsson's early Parabolic Trough Concentrator

Ericsson was no idle dreamer. Ericsson was world famous inventor and highly respected engineer. As the inventor of the Civil War's "Monitor," Ericsson may have saved the Union. During the Civil War's waining years the South unleash their secret weapon: the Merrimack. The Merrimack was the first Iron-Clad warship to take up arms against the

US Navy which were all traditional wooden sailing ships. The Merrimack unleashed devastation against the wooden war ships of the North. Imagine being on a Northern ship and firing your cannon against this "aberration," and having your shots bounce off!

In comes Ericsson. Filing hundreds of patents, designing and building the North's "Monitor," Ericsson may have saved the North. Had not the Monitor become victorious over the Merrimack the South could have blockaded the ports of the North with their Iron-clad super weapon and perhaps changed the outcome of the Civil war.

John Ericsson was the inventor of the "screw propeller." This invention made Ericsson famous worldwide for his propeller made steam powered river and ocean-going vessels possible and changed the world of transportation with powered ships.

The experience of the Civil War affected Ericsson, the terrible waste and destruction of lives. The horrible perverted folly of war changed this brilliant inventor engineer, and focused him to a new purpose - a solar powered Industrial world. After the Civil War, Ericsson began to see the world in a new light, a world which could live in Peace as the norm, not conflict.

A world of equity, self-reliance, and the vision that all the fruits of the Industrial revolution could be, and should be enjoyed by all peoples of the world.

A world of hope based on reality. After all, Ericsson was at heart an engineer - a practical man. Ericsson yearned to create a world of industrial peace.

Above view - Ericsson's Hot Air Heat-Engine

Ericsson knew to be truly free and self-determined you need a power supply. And, the only power supply of industrial strength distributed daily to everyone on earth is: the sun. A solar powered industrial world is logical, practical and being available to all peoples the most benign and most promotes the notion of personal and commercial freedom. The one industrial scenario where everyone, everywhere would have the essential energy needed to pump water, distill water, power steam engines for driving cottage industry.

Concentrated solar energy to be used to produce industrial power densities capable of moving industry.

Ericsson saw what Mouchot did decades earlier, and others would soon share. A passion and engineering practicality toward doing work with solar energy. A world bathed in energy available for anyone anywhere to tap.

The Industrial Sun became the obsession and focus of Ericsson's extreme brilliance and engineering talents, setting his sights publicly on Solar Power for the industrial world. Perhaps it's the loss and waste of war which sets a man of such engineering genius to turn to the question which can "obsolete" the very notion of war: energy available to everyone, everywhere. Power with no fuels, or fuel costs, no pollution, and no political or Military stress.

Ericsson wrote, "A great portion of our planet enjoys perpetual sunshine. The field therefore awaiting the application of the solar engine is almost beyond computation while the source of power is boundless. Who can foresee what influence an inexhaustible motive power will exercise on civilization and the capability of the earth to supply the wants of our race?"

Unfortunately, John Ericsson was an old man and passed away in the late 1880s. Taking his purpose to his death bed, it's reported his only regret "was, not bringing to the world a perfected solar engine."

Fortunately, other Solar Pioneers inspired by Mouchot, Ericsson, and their own hearts and minds soon picked up the mantle. As the late 1890s rushed towards the dawn of the 20th century - the energy wars were on. Who would win?

Would the 20th century be mostly powered by decentralized and self-sufficient Biofuels as Diesel advocated? Would the 20th century be solar powered as Mouchot and Ericsson demonstrated? Would electricity, which already displaced the entire "Kerosine" rock-oil lighting industry, be poised to be the future transportation system of choice?

The Electric car had many advantages over the early internal combustion engines.

The electric car didn't require you get out of the car and crank the starter. It was smokeless, and silent. Early automobiles were loud and smoky making electric cars more suitable for fine or crowded occasions, and certainly more suitable for women drivers Also, with an expanding electrical grid (keep in mind there are not many gas stations yet) the electric car could be charged-up many more places.

Battery technology had improved greatly with the Lead-acid battery which could be re-used many times and had a respectable power. Keep in mind U-boats were soon running underwater on Lead-acid batteries, so powering a car was not unimaginable.

By 1907 nearly one quarter of all police vehicles in NY City were electric, for example. However, the achilles heal was the lack of range, susceptibility to cold, and the charging rate of the batteries. Three issues modern Electric Cars have solved, but past 1910, the limited range of the Electric car was falling far behind the internal combustion engine with the great power density of gasoline.

By 1907, another threat to Rockefeller's empire of Kerosine now successfully transitioned to Gasoline production began peaking its nose under the tent: industrial solar power.

No doubt aware of Ericsson's work in solar energy engines, Rockefeller was pacified largely through the 1890s regarding solar energy's threat to his growing empire of "selling everyone fish everyday" through gasoline sales. But, working quietly through the 1890s emerged two other Solar Pioneers which gave Rockefeller cause to take another look.

The Southwest region of the United States, ie. California, Arizona, New Mexico has long been on everyone's radar for selling energy. The water pumping irrigation markets were unlimited as energy for pumping water could irrigate vast tracks of farm land. Short on readily available coal and natural gas resources in the Southwest, the early solar energy pioneers focused greatly on this region as a great prize. Irrigating sunny California for example was worth millions of dollars per year and a great fit for solar energy engines.

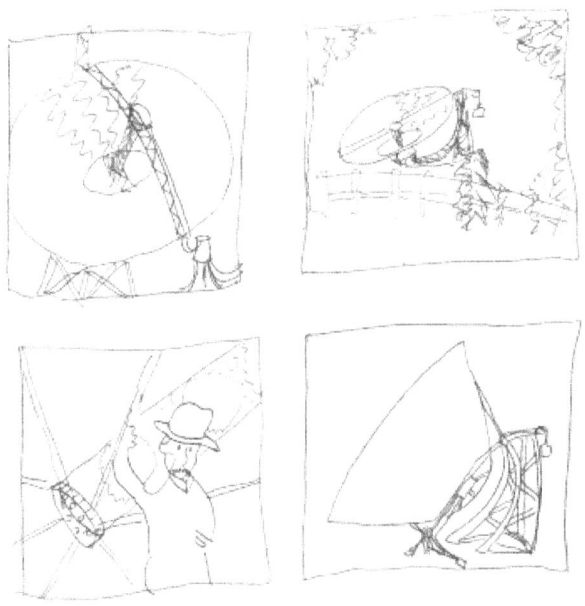

Shown Above - Enea's Solar Engine in Pasadena, California 1904

in 1892, an English inventor who emigrated to Massachusetts set out to improve upon Mouchot, and Ericsson's work and perfect the industrial solar engine.

In 1892, Aubrey Eneas along with some British investors formed the Solar Motor Company of Boston with the purpose of designing and building industrial solar engines. Living outside Boston Eneas immediately set to work studying all of the available literature and previous work on solar technology.

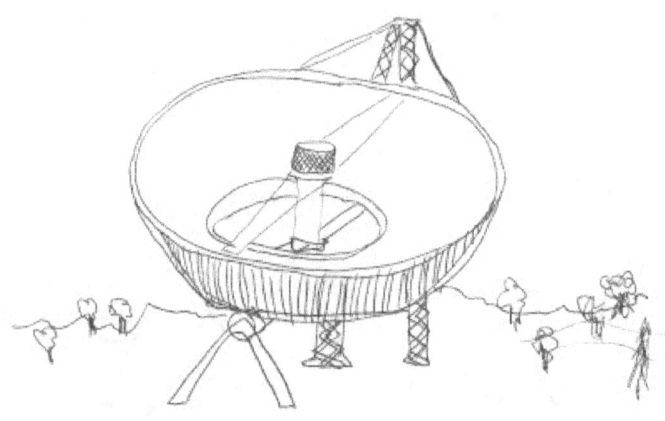

View above - Enea's Solar Engine irrigating an Orchard of 300 acres

By 1898, Eneas had demonstrated a working solar engine similar in design to Ericsson but twice as large. Ericsson had developed a low to the ground Parabolic Trough Concentrator (PTC) which reflected and concentrated sunlight upward from the reflector to an absorber pipe placed down the length of the collector.

Eneas realized his accomplishment but didn't yet achieve the 1,000 degrees F. temperatures he was seeking to power large steam engines. Back to the drawing board Eneas returns to Mouchot's original optics which resembled an inverted lamp shade.

The large open end turned to the sun with the interior walls using thin reflective Silver plates concentrated the solar energy evenly into the center line where Mouchot, and now Eneas placed his boiler. Eneas greatly increases the size of Mouchot's solar concentrator and replaces thin Silver plates (which tended to tarnish very quickly when exposed to the weather) with new inexpensive Plate Glass Mirrors as Ericsson had innovated.

By 1901, Eneas had achieved an impressive solar engine able to pump 1,000 Gallons per Minute! Outside of Boston this was an impressive feat. Eneas realized it's time to go out west to test his machine in the lands bathed in solar energy and thirsting for power for water pump irrigation. Eneas soon meets an old friend, Cawston, who happens to own the only Ostrich farm at the time in the United States. The location: sunny Pasadena, California.

They agree the Ostrich farm, where Cawston raised Ostrich for their plumage to supply the Ladies Hat market, and already a popular tourist attraction in

Southern California where Cawston sold tickets at the gate for people to tour the Ostrich farm - would be the perfect location to demonstrate Eneas's new Solar engine.

Cawston had a 300 acre orchard to be watered, and using Eneas's solar engine for irrigating the orchard would be the perfect demonstration site, and create more of a draw for tourists to the Ostrich farm.

The handbill read, "NO EXTRA CHARGE TO SEE THE SOLAR MOTOR - The only machine of its kind in the world in daily operation. Fifteen horsepower engine worked by the heat of the sun."

By 1903, Eneas had put it all together in a solar machine much larger than his Boston machines and with a mouth 32 feet in diameter concentrated solar energy to such intensities the machine pumped 1,400 Gallons per minute! To the amazement of everyone.

Reporters flocked to the site to see the amazing sun engine. F.B. Millard, one of the reporters to visit Cawston's Ostrich farm was so impressed he wrote, "... solar motors will before long be seen all over the desert as think as windmills in Holland, and that they will make the desert bloom as a rose - a phrase that literally represent the possibilities of the machine."

Typically, the solar engine would start pumping water about 1.5 hours after sunrise and continue for an hour or so after sunset. Millard describes the process:

"At first the morning dew is seen slowly to ascent in a wreath of vapour from the gigantic mouth. Then the bright glasses glitter in the morning sun, and the heatlines begin to quiver inside the circle, the greatest commotion being about the long, black boiler, which, as the intensity of the focused ray's increases, begins to glisten so that in any photograph taken of the machine, the boiler is shown almost pure white. Within an hour of turning the crank and getting the focus there is a jet of steam from the escape valve. The engineer moves the throttle, there is a succession of hisses from the boiler, a "clank-clankety-clank! - and the sun is drawing water in a way which he little dreamed a few months ago."

Another reporter who witnessed the specter so impressed wrote:

"Should a man climb upon the reflecting disk and cross it. He would literally be burned to a crisp in a few seconds. And a pole of wood thrust into the magic circle flames up like a match."

After two years of daily operation Eneas was ready for business. In 1904 he incorporated the Solar Motor Company in California and opened offices in the Bradbury Building in downtown Los Angeles. Eneas was humming now and had sales in Mesa, and Tempe Arizona.

Everything was beginning to look up when a series of disasters suddenly befall Eneas.

Despite the brilliant engineering of Eneas, he even designed and built a "clock-work" solar tracker which automatically tracked the sun throughout the day once the machine was positioned in the morning, he could not over come the power of severe storms.

A series of freak high wind storms and hail destroyed both machines' optical systems and put the brakes on his company. The relatively high optical profile of his concentrator added to wind loading stresses, and in severe weather, weakened by the endless heating and cooling cycles in the daily sun the metals in the frame weakened and caused within months of each other his two machines' optical systems to collapse during separate extreme storm events. By 1907, Eneas hits a financial wall and struggles to continue.

Another solar pioneer breaks onto the scene and achieves even more spectacular performance. Frank Shuman in 1906 enters the solar energy business with determination, engineering skill, and insight.

Shuman wrote, "One thing I feel sure of... and that is, that the human race must finally utilize direct sun power, or revert to barbarism." Shuman believed the one true hope for the future of mankind, as an industrial being, must be powered with solar technology. Setting himself to work Shuman reviewed all of the solar technology which came before and chose the best of everything he could find - the rest he would innovate.

Shuman begins with a one square foot "hot box" and rigs up an absorber which gets hot enough to power a small toy steam-engine he picked up in a toy shop for $1. Building rapidly on this Shuman, who lives just outside Philadelphia scales up his machine in the back yard and is soon achieving 5 horse-power.

Shuman innovates a "low temperature - low pressure" engine which uses the same principle when water boils at a lower temperature at altitude. Boil water in Denver and it boils at a lower temperature than at sea level. By isolating the internal pressure of the heat exchangers from the atmosphere, Shuman is able to power an engine pumping water using his new low-temperature, low-pressure steam engine.

Shuman continues to innovate combining hot-box technology with side reflectors and by 1911 Shuman's solar engine pumps over 1,000 Gallons per Minute to a height of 33 feet.

Shuman, with support from eager investors form the Sun Power Company and set off to transform the world. Shuman's investors insist on a large demonstration and they decide to build a large solar power plant in Egypt. A large plantation south of Cairo is selected, but before the large investment is raised, the board of directors want Shuman to go to Denver first, and build and test the machine before they ship it to Egypt, and for good measure the investors send out a famous physicist named C.V. Boys to review the project.

Boys points out under the current optics the bottom of the hot boxes are losing more heat than they gain and suggests a Parabolic Trough Concentrator (not unlike Ericsson 30 years earlier) which would irradiate a suspended hot box on the bottom and both sides as sunlight is reflected from the trough reflector up onto a long hot box arranged along the length of the solar concentrator.

Above view - Frank Shuman's Egypt Solar Power Plant, 1913

This improvement brings the output up to 4,000 Gallons per minute to the delight of everyone. Ready for Egypt the newly designed system is constructed on site and by 1913 the solar power plant is ready for demonstration. Shuman's new solar engine was incredible producing over 55 horse-power. You can imagine the Grand event with elegant ladies and men in their top hats and pith helmets touring this solar energy machine pumping over 6,000 Gallons per Minute! using no fuel and only sunlight in a desert bathed in sunlight as far as the eye could see - it was a new world. It's a solar industrial world.

Shuman became world famous and even Scientific American reported "a thoroughly practical machine in every way."

"Sun power is now a fact and no longer in the "beautiful possibility" stage... It will have a history something like aerial navigation. Up to twelve years ago it was a mere possibility and no practical man took it seriously. The Wrights made an "actual record" flight and thereafter developments were more rapid. We have made an "actual record" in sun power, and we also hope for quick developments." Frank Shuman, February 1914

The next thing to happen: a disaster. Within months of Shuman's grand demonstration of industrial solar power pumping 6,000 Gallons per minute! -using only rays of the sun ready to liberate the world - World War I breaks out, and all is lost.

Shuman returns to the United States at the war's outbreak and doesn't survive the war. Within only a few years the world lost two world-famous clean energy pioneers each achieving complete practicality and demonstrability. World War I is a disaster for clean energy as the world lost Rudolph

Diesel the leading world advocate for Biofuels, and Frank Shuman who demonstrated practical industrial strength solar power.

With the outbreak of World War I Rockefeller is able to secure major fuel contracts with the US Military, and he was active worldwide. The race was on to discover oil, control its sources, and use those resources to expand empire. World War I was all about empire.

Nation states were continuing the exploitation of peoples as it had been done for centuries - with force. Invade, wrangle control the countries natural resources, build railways only to those resources and export the wealth as quickly as possible. This world view has run the world of Empire for centuries, and with World War I the die is cast for the next century.

The 20th century would see incredible technological leaps in all fields, but almost all powered by rock-oils, coal, and other fossil fuels. It seems our modern technological power supply would be limited to fuels which can be sourced from the ground, controlled by men with guns, and sold piece meal to all who needed energy - everyone.

There was a moment in the early 20th century where clean energy could have been the path forward - Diesel and Shuman demonstrated its practical utility.

However, the death of these two pioneers sealed the fate of the world under the ambitions of such men as Rockefeller. Set against the ambition and designs of controlling the world's power supply for profit, the world had no chance against the tactics and ambitions of John D.

Rockefeller, and those who would be him. With the outbreak of World War I the Energy wars for the "fuel of the future" were over, with only one still standing: Rockefeller, and his business model of controlling world energy supplies for profit.

It would take a century for the world again to consider, is a fossil-fuel world really necessary? No, it's not. Not a century ago, and not today.

Solar energy is the only solution to industrial energy demand which needs no fuel, has no fuel costs or toxicity. Available everywhere to everyone, industrial solar energy as Shuman demonstrated is practical, profitable to the user, and causes no political or military stress.

The ancient Empires of the Bronze Age (Greece), and later the Iron-Age (Roman) were all powered with wood. Empires of conquest were assembled to insure its expanding supply. Over the last three centuries the industrial prime-mover has shifted. Coal replaced wood as the energy which fueled the furnaces of industrial empire, then came Oil in the 1890s.

Here in the 21st century with over 7 billion people on earth the stakes for human civilization are even higher.

Which path shall we take now? Will the 21st century pursue the Energy Empires which have flourished for millennia, or will mankind build a truly sustainable industrial future. based on a new paradigm?

Guaranteed poverty for billions, or guaranteed freedom for billions. The choice is stark. The Solar Water-Battery technology is the bridge to the latter.

Chapter Four: Toxic Time Bomb

The Fossil-fuel energy paradigm is based on a triangle of destruction. First, comes fuel costs. Energy is treated as a commodity extracting billions of dollars each day out of the world economy.

Second, when we consume Fossil-fuels we produce a whole host of toxins which impact our air, soils and water causing a wide range of deleterious impacts. Third, the inherent destabilizing effect of powering a civilization based on "holes in the ground surrounded by men with guns" supports wide spread corruption and extreme income inequity. It is a system which is inherently, and systemically unfair.

This chapter focuses on the second pillar of the triad of death: the Toxic Time bomb. There is a dirty little secret we must live up to, and this is the dark side of our civilization. By most accounts the world is nearly trashed. Every remote island in the world is awash in plastic and other contaminates. The gyres in the center of each of the great oceans hosts regions of plastic contamination boasting 100,000s of square miles in scope.

Plastics break up, but they don't break down. It's a toxic time bomb as each year unknown trillions of pieces of plastic breaks into two, then four, then eight pieces and so on. Every time plastic breaks up it increases its nefarious surface area. Breaking down into smaller and smaller pieces more and more wildlife cannot distinguish them from food. Birds are found starved to death with stomachs filled with plastic.

Fry, and fingerlings mistake them for diatoms or algae their natural food. As small fish eat these contaminates larger and larger fish, sea mammals, and birds who prey on them end up accumulating these toxins in their organs and flesh. This is bioaccumulation, and why its not recommended you eat more than two servings of deep sea fish per week. The oceans are polluted.

The danger of this toxic time bomb cannot be overstated. If the oceans die, we die.

If over fishing and plastics are not enough, the oceans are also under assault from our Coal-fired Power Plants. The massive amounts of daily emissions of CO_2, sulfates (acid rain), particulates, radiation and Mercury is having a devastating effect on our oceans. The massive amounts of CO_2 in the atmosphere are being in part absorbed by the oceans. Forming a weak Carbonic-acid when CO_2 and water mix an acidification of the ocean water occurs. This acidic toxic time bomb will only increase in danger and impact.

Crustaceans and other sea creatures which form shells do so in part from Calcium ions present in the water. Carbonic-acid formed from CO_2 dissolving into the water attacks Calcium ions dissolving them rendering them unavailable by ocean critters to form shells. The eventual outcome is yet another vector for absolute devastation for the health of the oceans.

Toxic time bombs as a direct result of the way in which we're industrialized as a civilization and are accumulating everywhere you look. Fossil-fuel use on such a global scale may have been a viable option three hundred years ago, but the toxic footprint we now empress on the natural world is accelerating unbridled and having an epic effect on the biology of earth.

The daily global emission of millions of tons of toxins worldwide has long been poising our soils, water ways and atmosphere towards a tipping point. How much degradation is considered acceptable? How much fouling is simply thought away as collateral damage?

This chapter is dedicated to the case that there is simply a limit to the amount of poisons and toxins any natural biological system can endure.

Since 1950, studies estimate there is now such a mass of plastics in the ocean it equals the mass of fish. It only took 66 years to accumulate such a large amount of plastic. What about the next half century? Will plastics literally choke the life out of the oceans and seas?

Basing plastics on Fossil-fuels produces most plastics which don't break down and exist for centuries, indeed acting as toxic sponges attracting other toxins afloat in the seas. Plastics which are in the water work as toxic concentrators by absorbing other toxins in addition to the ones they emit and therefore become a toxic time bomb for lower then higher life forms.

Competing for sunlight, how are natural diatoms and algae vital to the base of the oceanic food chain going to win this battle?

How can natural algae compete with a plastic competitor which nearly "doubles" in surface area, again and again, exponentially increasing in effect throughout future time?

The simple answer is: it can't.

At stake here and now in the early 21st century is a central question. Do we wish to survive? And, if we survive, do we wish a quality of life for the citizens of earth?

If we continue as a "business as usual" world then we guarantee a catastrophic biological collapse. The only question is when. Is it 5 years from now? 10 years? Perhaps, 20 years? The problem is the earth is facing catastrophic changes from man's current, and centuries old, method of civilization. The basic materials and energy we require are mostly taken out of the ground.

Sourcing our civilization's power supply from the ground creates all of these problems: high costs, increasing toxicity, and increasing inequity, political, and military turmoil.

If it were possible to solve all of these problems in one major leap, would we do it? If it were possible to make a list of all of the characteristics we'd like to see in an "Energy Paradigm" and solve for a solution, would we take a peak at how this energy world of the future would operate?

If I were to say there was a way to produce clean power anywhere on earth, with no fuel costs, no pollution, no toxicity, or moving parts, what would you say? I would further state this "power plant of the future" has no air, soil, or water impacts. Requires no combustion, no drilling, mining, fracking, strip-mining, pipelines, ocean tankers or trucks. Produces no Mercury, radiation, CO_2, Nitrates, Sulfates, particulates, VOCs or methane, if fact, has no emissions at all.

In a kernel, this "future power plant" has no fuel costs, no toxicity, no moving parts, can operate anywhere on earth, is safe, scalable, practical, sustainable, and just being a machine - affordable.

This "future power plant" is either a figment of the author's imagination, or humanity has been scammed into an endless cycle of poverty, abuse, militarism, and pollution destabilizing countries and peoples for the last one hundred years.

It finally must be said, "the Emperor has no clothes."

A world of 7 billion people cannot be afforded a "humane" existence if the power supply is based on Fossil-fuels and fissile material. In other words, a Fossil-fuel based world economy insures and indeed guarantees increased poverty, human suffering, environmental destruction, and political destabilization as the only future scenario - an amplified version of our world today.

If you wish to bring the entire population of the earth above the poverty level using Fossil-fuels - it cannot be done. It can't be done for three reasons: not enough Fossil-fuels exist, burning them would render the atmosphere and oceans uninhabitable, and the cost to do so far exceeds global GDP.

The simple fact is it can't be done with a Fossil-fuel paradigm. It is physically impossible to distribute fossil fuel energy in such quantities required for each individual on earth to live above the poverty line.

It's impossible, not to mention undesirable, to "distribute" fossil-fuels from where ever they are, to all of the people of the earth. In such massive quantities moving all this coal, oil, or and natural gas would have cost beyond the economic output of the world.

Let's apply the same standard and ask about this "future power plant." Can this "future power plant" provide enough industrial strength to power the electrical and chemical power demands of modern lifestyles, worldwide? Yes. Solar energy falling each day exceeds our demands 10,000 fold.

Thermodynamically, there is plenty of energy in the natural world. Solar energy powers almost all life and weather on earth. It can power industrial man too.

How does the 21st century power individuals, homes, communities, cities, and countries allowing not only a growth in commerce and economic activity, but doing so without choking ourselves to death with the filth of our own effluents?

Employing the Solar Water-Battery technology is the answer to this question. The 21st century industrial world will be powered by self-sufficiency using technology, or doomed to suffer the fate of the 20th century, and perhaps much worse.

Chapter Five - Why Sunlight? Why Water?

Why Sunlight?

The cornerstone of the argument for a solar powered world is - it's a solar powered world. Nearly the entire natural world from the perspective of earth is powered by the sun. This is no small feat. The earth receives each day over 10,000 times the total energy we consume for the entirety of our civilization for that day.

Solar power already drives all of the photosynthesis on earth producing the base nutrition which feeds almost all of the life on earth, and produces as a by-product the essential Oxygen without which all higher life forms would perish. It's no understatement to say that without the Plant Kingdom and its Solar power supply the earth doesn't eat, or breath. If solar energy powers the natural world, why not the industrial?

Indeed, all fossil fuels which we tap now for powering our civilization originated as ancient biomass produced originally from millions of years of photosynthesis powered by the sun. Fossil-fuels are a store house of ancient photosynthesis. All fossil-fuels began as solar energy.

The question at hand is what powers our Industrial Civilization historically, today, and what will power the future? Will the 21st century simply be an extension of the previous centuries, or will new conditions force humanity to make some stark choices?

In reality, it takes an enormous amount of energy, water, and resources to live as a human being with any level of dignity. As a 'back of the napkin' calculation we can begin to estimate the physical requirements for each human being, thermodynamically to live. A human life requires a minimum amount of clean air, clean food, and clean water to thrive, and each of these has an energy cost - not to mention any industrial needs of energy!

On earth there are living now more than 7 billion people. Once we calculate the minimum amount of energy required for each human being to live one day with a minimum of dignity we then multiply that energy value by 7 billion. Integrated over each year, it becomes clear it will take an enormous amount of energy, and in this scenario, available energy to everyone on earth to power a human civilization with clean food, water, air and economic activity.

Each of the great Solar Pioneers saw the only logical source of power to drive industry worldwide, available more or less to all people everywhere, must be a source of energy capable of industrial strength and tasks, as well as void of fuel costs, and toxicity. Only solar energy can meet these conditions. The solar pioneers shared this vision including Faraday, Mouchot, Ericsson, Eneas, Willie and Boyd, and Frank Shuman who wrote:

"I fear humanity must make direct use of solar power, or revert to barbarism."
                                          - Frank Shuman, 1913

Solar energy is indeed powerful, though often dismissed out of hand by commercial interests who would not have you learn how to fish.

Why Water?

Water is fundamental to both biological life and industrial life. Without water, we would shrivel in the sun and all life would perish. Our world is mostly covered with water, we are mostly made of water, we are a water world. Water is the most significant material on earth and it's vital to everything we are and everything we do. The special quality of water from an industrial perspective is its use as a feedstock for non-toxic energy.

Solar powered photosynthesis, the fundamental chemical process which enables almost all life on earth does so by achieving two chemical feats: the oxidation of water, and reduction of CO2 into fixed carbon molecules. As in all chemical reactions where one thing gives up electrons (Oxidation), and another thing receives electrons (Reduction). Chemistry always operates in "pairs" of these half-reactions and together form a chemical reaction.

Photosynthesis is a complex process using select bands of solar radiation to "oxidize" water breaking water into Hydrogen and Oxygen, and initiating a "cascade" of energy transfer into a complex series of reactions ending with the Calvin-Benson cycle which perform a "reduction" of atmospheric CO2 "fixed" into basic sugars - the building blocks of life.

As water is "oxidized" into Hydrogen and Oxygen, the Hydrogen is decomposed becoming a Proton and an Electron kept apart at just the right distance. These Electrons and Protons are held just far enough apart to form an "Electron Gradient" used by photosynthesis to propel charges needed further down the line in a long chain of reactions. The Oxygen is discarded.

We have an Oxygen atmosphere because 3.5 billion years ago Oxygenic Photosynthesis followed this path and ejects Oxygen as a waste product. And, today without the constant production of Oxygen from plants each solar day the earth would quickly suffocate.

Water is enormously valuable not just in biology and industry. Water's use as a feedstock for non-toxic rocket fuel is amazingly potent and was discovered 150 years ago unlocking one of nature's greatest secrets.

By 1839, Michael Faraday invented two machines and processes, among his many other inventions, which are vitally important. These two inventions are the Electrolyser, and the Fuel-Cell, respectively. The Electrolyser is a device which has two inputs: water, and electricity and two outputs: Hydrogen and Oxygen gases, respectively. The Electrolyser uses electricity to chemically disassociate the water molecule into Hydrogen gas and Oxygen gas stored separately.

Faraday, would recombine these gasses through his other invention the Fuel-Cell releasing a great deal of electricity, and producing the water back. Faraday discovered, invented, engineered, described, and produced devices which could decompose water storing energy, and releasing this energy back through a fuel cell restoring the water again.

This is the electro-chemical Water-Cycle. Water used this way becomes the "perfect" battery. Renewable energy goes into the water molecule making Hydrogen and Oxygen gases. Stored as separate gases they are available on demand to be recombined back into water through a fuel-cell releasing a high percentage of the electricity originally used to drive the process.

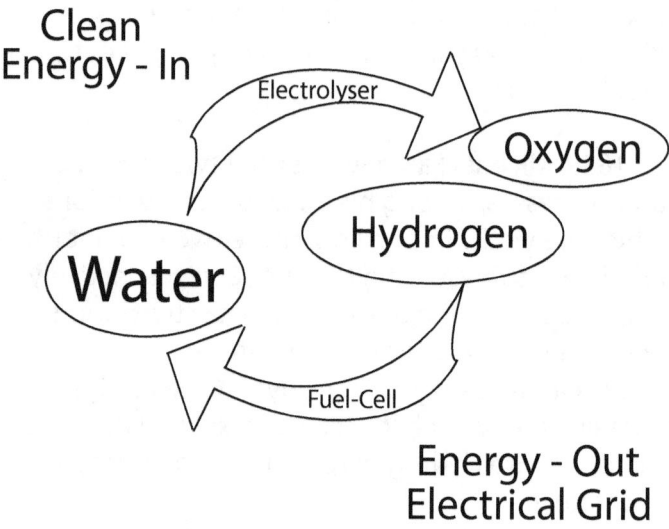

Clean
Energy - In

Electrolyser

Oxygen

Hydrogen

Water

Fuel-Cell

Energy - Out
Electrical Grid

Shown Above - The Water-Cycle

Storing gases as the perfect battery is relatively inexpensive, non-toxic, safe, endlessly reusable, works everywhere and has no emissions, air, water, or soil impacts. Water decomposed into gases can store energy in this manner available to power all industrial needs. Scalable, the Water technology can be configured to power any load from home to factory.

Michael Faraday points out circa. 1839 the ultimate fuel: Hydrogen, can be produced from a feedstock of water. Use any electrical input you like to drive the process. A few decades later Augustine Mouchot in Toures, France added the next step: power the process with the sun.

Why Hydrogen?

Hydrogen is the most abundant element in the known universe. From Hydrogen come all the other elements first in fusing to form the second most abundant element in the universe, Helium. Becoming in turn, through all the multiple star formations and sequences of stellar life and death - all other complex elements from which our existence is formed.

The third most abundant element in the universe is Oxygen. The universe has every reason to be considered "wet." Fortunate for us Hydrogen and Oxygen are the elements which together form water - without which we don't exist.

The earth is mostly covered with water, the human body is mostly made of water.

How amazing, the solution to our industrial needs is based on, wait for it - water.

What is the Ideal Fuel?

In earlier times I lived in Cambridge, Mass. One summer as a young man I afforded myself the use of the M.I.T. Barker Engineering Library under the big dome. A marvelous library where I found incredible scientific literature and journals from all around the world.

My quest: figure out - what was the ideal fuel?

My method was to compile all relevant information on all known chemical fuels. I proceeded to build a "Lotus 1-2-3" spreadsheet and began to look up the facts. I listed every fuel I could find, ... methane, propane, butane, ethanol, methanol, kerosine, gasoline ... the list went on. I noted in columns all the relevant factors including exothermic energy values during combustion in Kilo-calories per Mole (energy/mass). I listed and weighted how available the fuels were geographically, what their combustion products were, how toxic the products were, and on and on.

Finally, after two months of research I was ready to do my "Sort" and ask my questions. To begin, I needed to define what I meant by an "Ideal Fuel." I started with the question of Power. I figured the "Ideal Fuel" would certainly need to be powerful. After all, if you want to run my motor bike, factory, or country, you need a lot of power. At last I was able to highlight the column on exothermic energy under combustion - and pressed the button.

Whirling sounds from my disk drive indicated the program was engaged. After only a moment the top of the list stood blinking at me: Hydrogen. Hum, I remember thinking, that's interesting, and dutifully marked down in my notebook the result: HYDROGEN.

I thought about it a moment realizing, "that must be why they use it on the External Tank of the Space Shuttle, Hydrogen produces the most power produced per unit mass," interesting I noted.

Next, I asked, "the ideal fuel would certainly need to be clean," so I highlighted the Combustion Product Toxicity column, and pressed the button "SORT."

After the whirling, much to my astonishment at the top of the list again was blinking: HYDROGEN. Hydrogen! I exclaimed. Oh, I see. Burn Oxygen and you get Hydrogen Oxide. What is the chemical formula for Hydrogen Oxide? H_2O.

I was stunned by the logic, after all what could be cleaner to a planet covered with water, and humans mostly made of water, than water?

Again, I wrote in my notebook: HYDROGEN. Fascinated to continue I proceeded.

The next question I asked was, "well, the Ideal Fuel must be available," so I highlighted the column of Resource Availability and nervously pressed the key

"SORT." The whirling commenced and I still remember butterflies in my stomach. At last the top of the list stared blinking at me, as much as I was staring unblinking at it: HYDROGEN.

Of course, I thought, on a planet mostly covered with water, a perfect feedstock for Hydrogen is in the water all around us, inside us, falling on us, surrounding us, and being us. Water is the key to Hydrogen, which in turn, is the key to industrial society. I was floored.

I realized then. There is no real debate about which fuel is the Ideal Fuel. If your criteria is High Power, Non-toxicity, and near universal availability, there is only one answer: Hydrogen.

There is only one chemical compound, and in this case an element, which can satisfy all of these conditions simultaneously: Hydrogen. And specifically, Hydrogen fuel sourced from ordinary water. Make fuel from water, use the fuel, and get the water back.

Using renewable energy on the front end of this process and you get industrial sustainability. Therefore, Solar energy and water is the technology of the 21st century.

The only technology which satisfies the conditions of industrial strength power, non-toxicity, political neutrality, and add to that - no fuel costs.

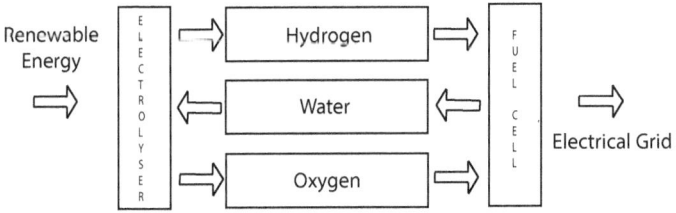

With a Solar-Water power system, as envisioned by Faraday and Mouchot, a worldwide power technology available to everyone, and everywhere can be established. An Energy paradigm can be engaged offering no moving parts, and no emissions. No air, soil, or water impacts. No radiation, or Mercury. No pollution of any kind. No need for mining, drilling, fracking, pipelines, trucking, ocean tankers. No need to pollute our oceans, no acidification, no VOCs, no particulates, no CO_2. In short, no problems which are the cornerstone and inescapable reality of life on earth in the early 21st century.

Our modern civilization has become so intoxicated by the industrial drug of fossil fuels we've become effectively "criminally negligent." For how can any sane civilization consent to destroy the very foundations upon which their children will soon depend? How can a sane civilization expand an industrial path with such toxicity that continued use threatens our own human species, and many other species with biological collapse?

The Emperor has no clothes.

A Call to Action

Understanding the Solar-Water technology pioneered by Faraday and Mouchot, there is no basis on which our world can rationally build another coal or nuclear power plant. Indeed, we need to systematically decommission and replace all current coal-fired and nuclear power plants around the world with Solar-Water Battery technology. Everyone gets their electricity. The difference is it's much less expensive, and produces no pollution, or motivation for war.

The worlds Energy Operating System has to be "upgraded" for the 21st century. A Power System 2.0. The old operating system of fossil fuels is no longer functional. The old fashioned notion of Heat-Engines powering Steam-Engines is now obsolete. There is no need for any major breakthrough. The root technology as it turns out is over 150 years old.

Today's chemical and media companies parrot the "joys and woes" of energy markets going up and down - far removed from the question of whether this market should even exist. "Oh, what a beautiful suit!" The business journals and TV news programs proclaim cheering the "low cost of oil" as it hits $30 per barrel as it did recently, yet lament the fact that billions of dollars were lost in world stock-markets evaporating over the first two months of 2016 during the same period.

When market prices for Oil go up, the world pays billions of dollars in increased fuel costs. When market prices for Oil go down, billions are lost in stock market value worldwide. Either way the market swings, as swinging is a market's nature, causes economic disruption. Either way the market moves, consumers (and shareholders) lose. Not too good of a system, eh? Unless you're at the receiving end of the Fossil-fuel business its a no win scenario for anyone.

Now, we live in the 21st century and a new day dawns. A day where the veil is lifted, and the reality about energy and empire can be looked at fully, literally, in the light of day. The choice is very clear. Either the future industrial world will be sustainable as achieved with Solar Water technology. Or, the world will continue to fall off the cliff of "un-sustainable," and loose everything through toxic collapse. At stake, is all that humanity has ever been, and - perhaps - all humanity may ever be. Those are big stakes.

An industrial world based on and powered by Fossil fuels only guarantees a future century just like the past - only more extreme. A world full of conflict, abuse, people fighting each other over energy, tapping only limited resources stashed away in the ground, controlled by those who would use them by guns and markets.

At last, this deadly foolishness can end. Let everyone learn how to fish for themselves becoming energy independent. Let's create new wealth. New wealth powered by a power supply which doesn't need to be "bought and sold" from suppliers to markets. Solar energy is already distributed. Let's tap new wealth powered by the source of power which, ironically, has always powered this world from the very beginning: the sun.

The sun powers the Natural world, why not the industrial?

The Water-Battery technology is simple in architecture and objective. Water is used as a feedstock and with clean energy electricity is chemically disassociated into separate streams of Hydrogen and Oxygen gas, respectively. The Water Battery enables the profitable use of clean energy by eliminating the variability issue.

The energy paradigm of self-sufficiency means energy independence. Energy independence is the basis of human independence and cannot happen the other way around. The Water Battery can be powered by any variable renewable energy inputs. Solar photovoltaics, wind generators, anaerobic digesters, tidal power; there is a long list. Every location on earth is favorable to one or more of these powerful inputs.

The Water Battery technology bridges the variability of clean energy inputs into a safe, reliable, industrial and potent power supply by providing energy on-demand with no toxicity. The Water-Battery stores these gases, separately in Cylinder Tanks. That's it. That's the universal storage technology: a pressure vessel known as a storage tank.

Electricity is produced on-demand by simply recombining these gases through a Fuel Cell. The Fuel-cell through an internal stack of carbon plates brings the gases together but uses a trick. Using a "proton exchange membrane" (PEM) the Fuel cell incorporates an internal barrier with "holes" only a Proton can get through.

The Fuel Cell trick is to make a wire connection (through an electric load you wish to power) providing an electrical path for the Electrons. As Hydrogen is fed in one side, and Oxygen the other side of the Fuel Cell, the Hydrogen can "see" the Oxygen and very much wants to reach the Oxygen side - prevented by the Proton Membrane barrier. After all, these two gases want to make water again and release their energy relaxing at last as stable water. The Fuel-Cell tricks the Hydrogen by having it "break-apart" with the Protons traveling through the PEM membrane to the Oxygen side, while the Electrons forced to "go around" and take the longer path through the wires (powering your electrical load along the way).

Finally, like a long overdue family reunion, at the end of the Fuel Cell where the Oxygen is fed in all three meet: the Protons through the membrane, the Electrons (after their journey through the wires doing work for us), and Oxygen all converge and happily reform back into water. Repeat. There are always some losses in a process and some water is consumed but it is a small fraction of total volume of water used in this process.

As long as you have energy (solar, wind, anaerobic digesters, wave, tidal or any other clean energy generator) you'll have a non-toxic, powerful battery to store it - until you want it. As soon as you need Electricity the Fuel-Cell engages and the stored clean energy is released forming the water again. That's it. No need for Drilling, Mining, Fracking or any other toxic emission or activities what so ever. The 21st century, can be powered industrially worldwide from Sunlight and Water.

If humanity wishes to survive the 21st century, then we need to engage a wide spread Clean Energy paradigm or face collapse. Biological collapse is guaranteed if we continue our fossil-fuel path. The only debate can be when - how soon?

The only logical solution to distributed world power generation is to use Renewable Energy on the front end with the Water-Battery as the fundamental storage and power conversion architecture. There is no other physical energy technology which satisfies the conditions of no fuel costs, no toxicity, no emissions, no air, water or soil impacts, and as importantly no political or military advantage or disadvantage.

There is a "Silver Bullet" if you will. It's based on technology 150 years old. There is no need for ground-breaking research, there is no need to "hope" for some distant future when someone "invents" a breakthrough which allows all people to live in ideals. The breakthrough was done 150 years ago.

The "decision" not to engage it is purposefully controlled by Chemical companies and countries' interests who make a lot of money selling you fish. That's it. It's nothing more than making fists of money, while denying others the ability to live in their own freedom. The Fossil-fuel world only works if you have a "have" and "have-not" situation. Rockefeller was the grand architect in the modern age of this economic energy paradigm. Unfortunately, the 21st century can no longer run on this charade.

The entire global energy system of fossil-fuels guarantees, and is the economic basis of human poverty. Poverty can be rendered obsolete in a solar powered industrial economy. The Water-Battery makes the notion of human poverty absurd, and cruel - its true face. The "myths" about solar energy are pursued by those who would sell you fish, and their collective interests have sabotaged media and knowledge through every institution.

The time has come in human affairs to recognize a "Declaration of Human Independence" which values, supports and Power's a 21st century global citizen's right to energy, to history, biodiversity, environmental integrity, and most importantly, to individual human dignity and the dignity of others in this life.

The Solar Water-Battery technology is a technological declaration of human independence with an eye towards our children's world being more prosperous, less toxic, offering more true freedom and independence for anyone living anywhere than came before. The Fossil-fuel game is up. The Emperor has no clothes.

The Clean Energy revolution is profitable to use, sustainable, and industrially smart. Everyone wins. Everyone can fish. And, our children's children can live in a world of opportunity, appreciation, reverence for the Earth, and a humanity which expresses and lives to heights yet undreamed.

A human civilization where dumb poverty, and the suffering of billions is a distant memory, where the notion of systematically polluting our oceans, air, soil and water is discarded. A human civilization everywhere energy-rich, where the full potential of the human spirit is free to grow and to express.

The Solar Industrial revolution is a human revolution as revolutionary to our species as fire was so many millennia ago. With a real power supply available to all humans, there is no need to want. There is every reason: to be.

Chapter Six - Sunlight and the Water Battery

Renewable Energy

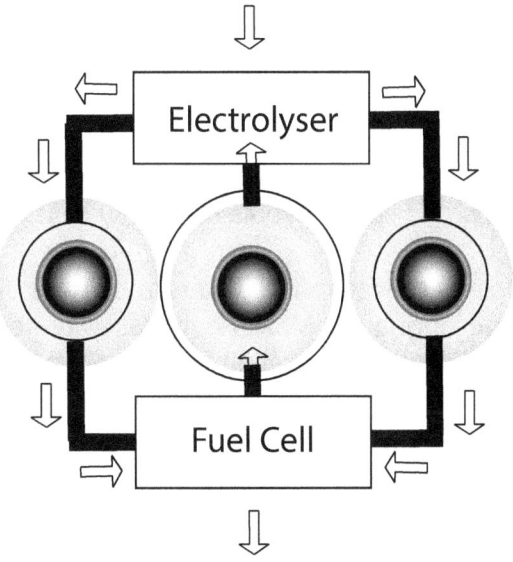

A modern Industrial world requires a lot of energy. People need this energy in cities, on farms, while traveling in any manner of vehicle. We need power for heating and cooling our homes and buildings, and we need energy to power industry and agriculture.

For over three centuries the enormous energy stored in Fossil-fuels has powered our machines at great cost. Great expense in direct fuel costs, great expense in dealing with increased toxicity, and great cost in the political and military instability caused when some have resources needed by all.

A Fossil-fuel world has concentrated enormous wealth into a few hands which in competition with each other encourages military adventurism and political destabilization each with enormous price tags generating enormous profit for a few.

All of this is because of a ruse. "Why teach people how to fish, if I can sell you a fish every day?"

This really is the reason. If it weren't so tragic, it would be laughable. Make no mistake, we get a lot from our Fossil-fuels, after all, it's concentrated Carbon and burns hot and powerful, however partially.

The problem has been accumulating and building its own inertia such that the fuel costs, toxicity, and the political collateral damage is becoming insurmountable. Something is missing.

The Solar Water-Battery Power Technology - How it works.

The Solar Water-Battery power technology begins with a tank of water. This tank of water separates a tank of Hydrogen from a tank of Oxygen. In front of this assembly is placed an "electrolyser" which is connected to all three tanks. Behind this assembly is placed a Fuel Cell which is also connected to all three tanks.

In short, water is decomposed into gasses in the first step, stored, and brought together in a fuel-cell to produce electricity on-demand with the water back in the second step.

The electrolyser has two inputs: electricity from renewables, and water from our center tank. The electrolyser decomposes water into its constituent gases Hydrogen and Oxygen, respectively. Under pressure these gases are pumped into their respective tanks. This is the battery. Stored gases. As a battery it's amazingly advanced, stable, potent, and safe.

Hardly any moving parts, just gases through pipes and occasional valves opening and closing. The stored gas battery can be "cycled" effectively forever, where all typical chemical batteries have a limited lifetime requiring replacement after approximately 2,000 cycles.

As stored gas there are no chemicals involved so nothing to wear out or be toxic. Energy stored separately as Oxygen and Hydrogen can last decades and provides the highest power density per mass than any another other known fuel or oxidant combination.

Described as a "flow-type" battery as long as you have solar or other renewables on the front end producing electricity and water from the water tank, vast amounts of energy can be stored safely for use at anytime, reforming the water back when used.

To produce electricity on demand the Fuel-Cell stack recombines the Hydrogen and Oxygen gases inside the Fuel Cell. This process releases almost all of the energy stored originally, small amounts of heat, and recycles the water back. Yes, we get the water back (most of it typically over 99.99%)

This Water-cycle battery power converter is powered on the front-end by electricity from renewables such as Solar PV, wind generators, anaerobic digesters or other clean energy sources producing electricity though variable with no fuel costs, no toxicity, and no political stress - as this process works anywhere on earth.

The Solar-Water power plant is an ideal replacement for our base-load electrical power plants such as Coal and Nuclear power plants. The Water-Battery allows the input of any clean energy source with variability rendered inconsequential. The Water-

Battery converts Variable inputs into Steady on-demand electrical outputs.

From an architectural perspective all electrical power plants essentially have three steps: energy input, energy processing, and energy output, respectively. Fossil-fuel power plants use Coal, Oil, or Natural Gas as the "energy in" bit.

Energy in from Fossil-fuels required drilling, fracking, or coal mining and include transportation, storage and stockpiling the fuel to later burn to boil water driving a steam turbine, or in the case of Natural Gas a gas turbine. The "energy-out" refers to the power conditioning required to convert the electricity into suitable voltages for transmission and distribution to an end-consumer.

In the Fossil-fuel model the energy-in and processing part is vast. Mining, fracking, drilling, railroads, trucks, tankers, strip-mining are all involved moving Carbon from where its dug up or drilled, to where it's ultimately burned to power a heat-engine.

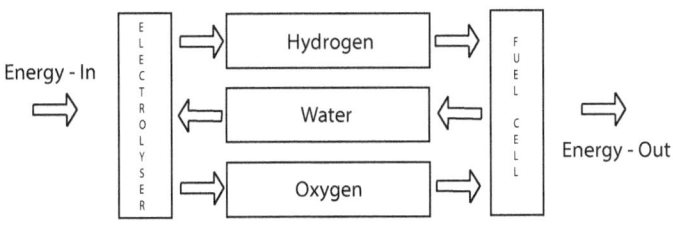

Above - The Water Battery

Using the Water Battery the total fossil-fuel world can be replaced by solar photovoltaic (PV) panels which produce electricity using a solid-state method with no emissions or moving parts as the front end of the power plant. (Including, other renewables ie. wind, digesters, tidal)

Energy processing using the Water-Battery replaces all of the Fossil-fuel impacts by starting with a safe and stable feedstock: water. Faraday, in 1839, demonstrated his Electrolyser device, through which water can be electro-chemically decomposed into Hydrogen gas and Oxygen gas stored separately.

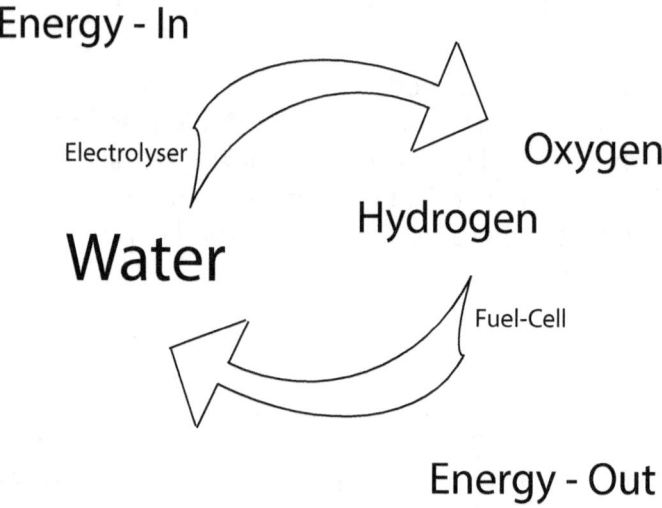

Energy - In

Electrolyser

Oxygen

Hydrogen

Water

Fuel-Cell

Energy - Out

Faraday completes the Water-Cycle by inventing the Fuel Cell which recombines Hydrogen gas and

Oxygen gas, respectively back into water - releasing a great deal of energy. Thats it. There is no magic, no mystery, no massive funding required. It's just building on what science has known for 150 years.

Use clean energy to "charge" the Water-Battery. Use Fuel-cells (the water battery) to produce electricity and repeat.

There is no toxicity. The Water-Battery engineered properly is safe, scalable, powerful, and can operate in any location on earth. Powered by fuel-cost free renewables the "fuel-cost" economic component is removed. The toxicity component is removed. The political instability is removed, as everyone everywhere can engage this technology.

The Water Battery makes it possible to store residential, commercial, and industrial scales of power production by scaling the system. That's the point.

The Water Battery makes possible a new industrial age which can actually bring the populations of earth above the poverty line. Human civilization, to be able to afford dignity to its citizens in the 21st century must command great energies.

For the last three hundred years fossil fuels have been the "prime mover" and the fuel of the industrial age. However, the needs of the 21st century are far more acute than centuries before and human civilization has "out-grown" the simple notion of burning rock and rock-oils to power the world. Humanity must evolve away from "burning" things.

The Water-Battery with solar and wind inputs, or any other clean energy generator makes possible the "retirement" of centuries old technology, ambitions, and profit motives. The Water Battery is the "missing link" which has been more hidden than missing to protect commerce.

With 7 billion human souls alive on earth the time has come to release the true human potential and propel human civilization into the 21st century which has been the dream and purpose of so many human lives through history: the creation of a world of dignity, fairness, and freedom of expression.

The freedom to live in happiness, while in the pursuit of happiness. A power supply for Industrial Man. The only thing separating a bright future for mankind, from total disaster hinges on its power supply.

The Solar Water-Battery system is the power supply for that journey, bridging the Fossil-fuel past with the Clean Energy future.

Chapter Seven - The Clean Energy Economy

Today, the economics of fossil-fuel energy is based on a Commodity. If you need energy, whether a Kilowatt-hour of electricity, or a gallon of gasoline, as a commodity, you pay for it. Repeat.

Renewable energy has a systemic advantage over fossil-fuels by having no fuel cost component. Of course there are hardware costs, but this is a fixed cost, not a variable and repeating fuel cost. The Water-Battery is a disruptive technology which challenges the paradigm of "pay as you go" to pay only once. Changing the economics of global energy from a Commodity based model into a self-sufficiency model is the challenge facing the 21st century.

The Fossil-fuel industry has for over a century operated under the motto "why teach you how to fish, when I can sell you a fish everyday?" This really is what it's about.

The Solar Water-Battery teaches people how to fish.

Clean energy is the most vital evolution we can engage in as an industrial society if the welfare of future generations matter. Failing clean energy, all indicators and major trends of toxic bioaccumulation show a mounting and impending cross over-point in the near-future where toxicity will cause a series of rapid collapses - starting with fundamental life forms at the base of the food web.

A series of toxic time bombs is building up from bio-accumulation until a tipping point is reached and biological collapse happens suddenly and irreversibly.

As a civilization composed of individuals, families, communities and countries we must come to grips with the importance of un-sustainability, and the nature of world economy. In all of the diversity of the Human race, we all share one inseparable fact: we're each human.

Economy, as any great tree has major branches. In our case the major branches are the major activities of our modern economic activity such as Manufacturing, Transportation, Communications, and Agriculture. However, it's the root of the tree which matters most in times of trouble. If your roots are strong then you can endure all kinds of stress and still recover. If your roots are diseased or weak, then the prognosis of the tree is at best - problematic.

The root of our "economic" tree for three centuries has been fossil-fuels. The problem is with the needs of the 21st century being much greater than ever faced by human civilization before, the call for action could never be more vital.

The Solar Water-Battery changes the world economic paradigm which has driven human affairs for 10,000 years. Classic economic thought defines a transaction as a market between "supply" and "demand." In classical economic models the "market" sets the price.

The problem with this thinking is it presupposes you need an outside party. Supply and demand presupposes you are the "demand" and that you need to seek out, negotiate with, and pay some outside entity for "supply."

In the case of Energy this has certainly been true for centuries. However, the 21st century with the Water Battery challenges this world-view by pointing out, with the right technology, you are both the "Demand" and the "Supply."

In the economic paradigm of the Water-Battery technology you are the supply. The entire notion of energy as a commodity becomes obsolete and absurd. Once you acquire the right Hardware, there is no additional costs outside modest maintenance. No fuel costs, no liability for toxicity, and no motive for anyone to invade you to take yours, when they can simply have one too.

The world economy would be transformed from a commodity high stress exploitative model into a distributed self-sufficient paradigm. A self-sufficient world energy model which is stable, profitable for everyone because their energy bills are greatly reduced, no unintended consequences with pollution so no down-stream liabilities, and no political or social stress. There is no longer a world of "Haves" and "Have-nots." It becomes a world of all Haves. Everyone can make their own pie, the notion of competing for one pie is - obsolete.

The fact is this earth is vastly wealthy. The natural wealth of the earth is incalculable. The tragedy is humanity has been delegated to being a "consumer" and lead to believe we must all compete for resources, and pay someone for the basic energies required for life, free expression and industrial expansion. The world is thought of economically as a zero-sum game meaning there will always be Winners and Losers. Someone has to win, and by definition someone has to lose.

In fact, there is more energy falling on a small fraction of our lands to accomplish everything you'd ever wish to do. The truth is there is more than enough energy for everyone on earth to live an energy-rich lifestyle - if you do it right. The poverty suffered by billions is a human institution, not a thermodynamic necessity.

One of the big "myths" perpetuated by the Petrol-chemical and coal companies throughout the 20th century was "without fossil-fuels, we'd all live in like cavemen!"

Of course, this is absurd. The solar Water-Battery allows anyone, anywhere to live as energy rich as they scale the system. There is more than enough energy falling on relatively small areas, for example the area of your house roof to power local electrical loads.

Hundreds of thousands of homes are totally powered by solar photovoltaics. Completely. It's been done thousands and thousands of times. To argue solar energy is "too dilute" for practical use is ridiculous. It's been done for decades.

The solar Water-Battery takes this concept to its logical conclusion.

How elegant, the industrial answer to world energy supply is based on what the world is mostly covered with, and your body is mostly made of: water.

How much is sunlight worth?

The earth receives trillions of dollars of wealth Each Day from the sun. How much is sunlight actually worth for human industrialization? Let's ask a specific question for perspective.

How much is solar energy worth falling on one-square mile of land per year?

Imagine you're a land owner in Arizona and you have one-square mile of land to lease. I approach you and say "Hey, I bet I can mine $150 Million - per year - of Gold off your land."

Your reply may be "Well, It's mostly sand and scrub brush, I can tell you for sure there is not $150 Million worth of gold - each year - on that land."

My reply, "Not all Gold is buried."

Let's do a calculation. To compare "apples to apples" let's define the price of electricity at 20 cents per Kilowatt-hour (kWh). One square-meter (m2) of land has at peak 1,000 watts of optical energy coming from the sun. Converting to the metric system one square-mile is approximately 2.5 million sq. meters (m2) in total area.

One sq. mile receives about (1,000 watts solar power x 2.5 million m2) which is 2.5 Million Kilo-watts (Kw) of optical power in each peak moment. Arizona has on average about 5.5 Peak solar hours of daily sunlight. Let's calculate the energy available over an average day. Multiplying 5.5 hours/day x 2.5 Million Kw equals 13.75 Million kWhs of Optical energy production - each day.

Now, to answer the question how much is sunlight worth? To compare Apples to Apples we need to convert solar energy into electricity.

Solar photovoltaic panels convert about 15% of solar energy into direct current (DC) electricity. Taking 15% of 13.75 Million kWh of energy yields just over 2 Million kWh of electrical energy per Day.

At 20 cents per kWh, an average electrical rate but put in any figure you wish, 2 Million kWhs of electricity is worth approximately $400,000 per Day! As a practical matter this calculation ignores packing density, but arrives at a number you can derate. In general terms it illustrates the enormous wealth the "raw material" of solar energy represents from an industrial power supply perspective.

One square mile of average land in Arizona receives about a $146 Million Dollars worth of energy each year.

This is the power supply which is the foundation of the clean energy revolution. Solar energy is real wealth. Sunlight is an industrial prime mover, if employed to be so. In nature Solar Energy is the prime mover, and for mankind's industrial civilization provides an ideal solution, and real power supply for general peace and prosperity worldwide.

The Clean Energy economy is based on Solar Energy's intrinsic value. Using modern Solar, Wind, Tidal, Anaerobic Digesters and other renewable technologies to produce the initial energy the Water-Battery provides the missing link to safely store these energies for industrial use on-demand 24/7, day or night in any season.

The Economics of the 21st century will evolve in essential sectors such as energy from a purely "supply" and "demand" transaction model, into a self-sufficiency paradigm tapping into and creating new wealth worldwide each day.

A future which provides enough energy to supply the needs, and indeed desires of 7 billion individuals worldwide ending the cruel and outdated notion of human poverty.

The Solar Water-Battery technology makes this future possible.

Mankind can ignore nature, or embrace nature. Embracing natural energies for the industrial uses of man provides a future and world where our children's children can thrive.

Chapter Eight- Thermodynamics of Peace

Throughout human history the story of Energy and Empire is really the same old story of "man gets stick, man uses stick, man needs more sticks, man needs more sticks to get more sticks."

Since the last Ice-Age ended 10,000 year ago and mankind moved into the valleys and discovered the advantages of being agrarian, a dark side was also unleashed beginning innocently as trade, and soon ending in conflict and war.

Looking at the earths resources as a limited pie, humans have historically embraced the notion of empire to ensure the influx of goods and services, wealth in short, to fuel the empires expansion domestically and abroad. Is our modern world any different in form and function from any of the empires which have lived and died over the last ten thousand years?

Is it possible to think industrially in a new way? To realize the earth is not a limited pie to be fought over? Using solar energy directly for industrial purposes changes this world-view by making it possible for everyone, everywhere on earth to make New Pies. To make New-Pies each day, which is to create new wealth because the sun rises every day.

The thermodynamics of peace would allow any human individual or group of humans living anywhere to provide self-sufficient industrial strength energy on demand using only renewable energy inputs, and water.

Any scenario envisioning worldwide general peace would certainly need to raise the universal standard of living of all people above the poverty level at a minimum. After all, how could there ever be peace in the world as a practical matter if children are dying in front of their parents? To actually raise the entirety of humanity above the poverty level takes a distributed power supply.

To achieve actual general peace in the world requires energy - a lot of energy. The only way this can be done under the conditions of no fuel costs, no toxicity, widely available, with no emissions or environmental impacts what so ever, is from Solar-Water technology.

It is often very popular to hear from so called energy experts "Ya know, there is no Silver Bullet" when discussing energy or climate change issues. They only believe there is no Silver Bullet, because to do so requires hard research, hard work and intent. The fact is, "there is no Silver Bullet" is the go-to line for all of the energy interests to squeeze in their consideration and blow-off criticism and justify their existence. "Well, everyone knows there's no Silver Bullet."

Poppycock. The author suggests they can't find a Silver Bullet because they don't seek one. Remember, you can't get answers better than the questions you ask. If we can solve 90% of our problem with the Solar-Water technology, well, in my book that's a Silver Bullet. It's time to coordinate worldwide and launch this Silver Bullet and transform the world's energy paradigm from a dirty consumer, burner, polluter and squanderer of resources, into a truly sustainable, just and toxic-free method of industrial power generation - accessible by all people and all nations - not in some distant future fantasy, but immediately.

The world can jump from fuel costs, toxicity and political instability to an energy paradigm with no fuel costs, no toxicity, and no political impact. The exact opposite. Our present Fossil-fuel paradigm is exactly upside down.

An Industrial Revolution is about to occur on earth where a century of misdirection and war, can at long last be set right again and humanity can once again pursue a path of culture, tolerance, unfettered expression, and joy in the diversity and sanctity of our wonderful earth. Thousands of human generations stretching back into deep time tell so many stories of heroism, great sacrifice, incredible love, and the unwavering desire to see the children of the world see a greater day than this one.

The 21st century will be the moment when evolution becomes revolution as a great leap is taken - away from the abyss of toxicity, death, pollution, corruption, environmental suicide and perpetual poverty, into an industrial world which lives in the light, literally, the sunlight which has powered the earth from the beginning.

In the last hundred years, human civilization has achieved technological accomplishments hardly dreamed of for millennia if ever dared dreamed by man. We can fly to the moon, live underwater, travel vast distances in hours, speak to each other in real time separated by continents. Our world technology, to anyone brought from history would be indistinguishable from magic. There is no way any human person, somehow transported to our modern world, could manage to understand or absorb the incredible "magic" and shear splendor of our technology - yet, we still "burn" things to power heat-engines to drive steam-engines and pistons. We, as a civilization, seem to be stuck in First gear. We're stuck in the energy paradigm of combustion, and sourcing our industrial fuels from the ground.

The Solar-Water technology of Electrical Power Generation is conceived and designed to achieve a fundamental goal: power on demand, energy accessible to all people with no political stress or military involvements or motivations, and with no toxicity to harm current or future inhabitants.

A physical anthropologist noted in a lecture, "Evolution is not a step-by-step process. The anthropological record shows: step-step-Leap. Step, step, step- leap. It's in the great discontinuities that evolution truly lives. It's not in the Steps, it's in the Leaps!"

In the 21st century human civilization meets its greatest collective crisis. Our energy crisis. Our world cannot achieve any real peace powered by Fossil-fuels. The 20th century proved this, with the beginning of the 21st even more violent. A technological leap must occur where the old paradigm of fuel costs, toxicity and instability are replaced with no fuel costs, no toxicity, and a power supply technology available to all human beings everywhere, relatively all at once.

The industrial future of humanity is at stake, uncounted millions of species and humanity face a toxic time bomb which if not diffused, dismantled, and rendered obsolete, will precipitate the certain untimely collapse of modern civilization in the 21st century.

Or, we begin systematically decommissioning all Nuclear and Coal-fired power plants worldwide producing many new jobs over this decade and replacing them with safe, powerful, renewable and industrially sustainable energy technology using the non-toxic Solar Water-Battery which provides a real means for everyone everywhere to participate in the marvels of 21st century technology, knowing the children of the earth, and their children have a brighter future than today.

Chapter Nine - Epilogue:
Electricity Market Opportunity

The world market for total energy exceeds $10 billion per day. Add up everyone's electricity bill, natural gas, coal, oil and gasoline bill an it's enormous. In rough terms, about 15% of the worlds GDP is spent paying someone for energy.

The Solar-Water energy paradigm changes all of this by offering the world no fuel costs, no toxicity, and no political inequity. Since this isn't magic, and is composed as a machine there is a fixed cost. However, no fuel costs and little maintenance costs keep this fixed cost easily managed, predictable, and structured into an amortized payment significantly below a variable commodity based market as is now the case.

Traditional energy markets use traditional power plant architecture. In the case of electricity a centralized model is used where a remote generator such as a coal power plant is located far away from a population to minimize pollution impacts to the market. (Out of sight, out of mind). The centralized model then transports the energy usually by High Power Transmission lines usually called T&D (Transmission and Distribution) to the energy consumer.

Your electric bill is not just for the Kilowatt-hours (kWh) you consume monthly, the electric utilities have recently added a cost structure known as "demand-charge." The demand-charge is based as a threshold of Power consumption at any one moment. For example, if the demand-charge is set at anything above a consumption of 20 Kilowatts (Kw), then the demand-charge is triggered and you're assessed an additional fee.

Over the last decade electric utilities in the US have grown in service areas about 10%. However, income has risen over 50%. The electric utilities have added the demand-charge to increase revenue. Auspiciously, under the notion of incentivizing lowering demand across a grid during peak times. LA in the summer, or any other service area, can have demand peaking when everyone has their air-conditioner turned on around mid-day.

The demand-charge was designed to apply "market forces" using higher prices to lower demand on the network during peak use. The Solar Water-Battery technology changes this notion by creating not a "Centralized" model of power generation and distribution, but more "Decentralized" mini-grids.

The electrical grid in the US is mostly three major grids: Western, Central, and Eastern. This centralized production then distribution model integrates multiple generators and consumers all balanced and wheeling power back and forth to meet demand. Electrical grids are very dynamic. When you flip a switch that load is sensed by the power generator and must increase its output in real time. The dynamic environment of these mega grids is truly amazing.

However, mega grids are also very vulnerable, and increasingly expensive to the consumer. The Solar Water-Battery mini grids allow you to power homes, communities, counties, municipalities and countries by using systems singularly or in parallel. Scalable and interconnectable the Solar Water-Battery greatly simplifies the nature of the electrical infrastructure, which has been been built over the last 100 years, using old centralized engineering. The new Smart-grid meets Smart power production.

The economic opportunity is enormous. Such a "Jobs Bill" has never been proposed which would seek to decommission all Nuclear, and Coal-fired power plants worldwide to be replaced with Solar

Water-Battery technology producing a net increase in employment.

As it was at the dawn of the 20th century, we have choices at the first part of the 21st century. Choices which can transform mankind's civilization into a world of increasing freedom and dignity, or, the continued free-fall now experienced by billions of souls on earth into toxic collapse.

The fork in the road is very clear. Continue Fossil-fuels and billions of people are condemned to un-ending poverty. Take the Solar Water-Battery path, and humanity can make a leap in evolution embracing a sustainable industrial society. A world where conflict has been relegated to soccer matches and chess tournaments . A future where a human declaration of independence applies to all people, and ultimately all species as we become in partnership with nature, and not at war.

The Water Battery technology with a clean energy front-end offers all the energy our modern civilization demands, with none of the problems our progeny deserves. The 21st century will set the industrial stage for the future of our human species. Will we survive the 21st century? Will the oceans? Only if we do the right thing and adopt Water Battery technology.

As true a hundred years ago as today, this new century is yet to be written.

About the Author

Christopher (Toby) Kinkaid

Christopher (Toby) Kinkaid, from Portland, Oregon is the founder of Solardyne.com, SolarQuote.com, and AlgaeToday.com, and has worked in the clean energy field for over three decades. Mr. Kinkaid is the inventor of the "**Helyx**" Vertical Axis Wind Generator, the "**Mariposa**" Non-imaging solar concentrator PV module (continuous operation at Sandia National Laboratory since 1994), the **Solar Demultiplexer** optical solar concentrating lens (Dr. James/Sandia National Laboratory 1991), and the inventor of the original "**Solar Power Pack**" (Mother Earth News, "**Littlest Utility**" June/July, 2001). Mr. Kinkaid has been an official lecturer and presenter on clean energy technology around the world including APEC, Bangkok, Thailand, 2003, "Energy

Solutions World", Tokyo, Japan, 2003, the International Biomass Conference (IBC), 2010, Minneapolis, MN, and the Algal Biomass Organization (ABO) Conference, 2010, Phoenix, AZ.

Mr. Kinkaid has appeared in TV interviews on KOIN TV, KGW TV, FOX 12 NEWS and "Sustainable Today" produced in Oregon. Mr. Kinkaid has served on the board of directors for the National Hydrogen Association, in Washington D.C., 1993, and the Japanese Satellite Communications Company (JCNET), Fukuoka, Japan, 1994, Oregon Wind Corporation 2003-2007, Algaedyne Corporation (2008-2011), currently serving as CEO of Solardyne in Portland, Oregon.

Christopher (Toby) Kinkaid is based in Portland, Oregon, and continues his work in clean energy technology and applications.